Fluid Mechanics and Its Applications

Founding Editor
René Moreau

Volume 144

The purpose of this series is to focus on subjects in which fluid mechanics plays a fundamental role. As well as the more traditional applications of aeronautics, hydraulics, heat and mass transfer etc., books will be published dealing with topics, which are currently in a state of rapid development, such as turbulence, suspensions and multiphase fluids, super and hypersonic flows and numerical modelling techniques. It is a widely held view that it is the interdisciplinary subjects that will receive intense scientific attention, bringing them to the forefront of technological advancement. Fluids have the ability to transport matter and its properties as well as transmit force, therefore fluid mechanics is a subject that is particulary open to cross fertilisation with other sciences and disciplines of engineering. The subject of fluid mechanics will be highly relevant in such domains as chemical, metallurgical, biological and ecological engineering. This series is particularly open to such new multidisciplinary domains. The median level of presentation is the first year graduate student. Some texts are monographs defining the current state of a field; others are accessible to final year undergraduates; but essentially the emphasis is on readability and clarity.

Springer and Professor Thess welcome book ideas from authors. Potential authors who wish to submit a book proposal should contact Dr. Mayra Castro, Senior Editor, Springer Heidelberg, e-mail: mayra.castro@springer.com

Indexed by SCOPUS, EBSCO Discovery Service, OCLC, ProQuest Summon, Google Scholar and SpringerLink

Lewis Johns • Ranga Narayanan

The Shape and the Stability of Pendent Drops

Springer

Lewis Johns
Department of Chemical Engineering
University of Florida
Gainesville, FL, USA

Ranga Narayanan
Department of Chemical Engineering
University of Florida
Gainesville, FL, USA

ISSN 0926-5112 ISSN 2215-0056 (electronic)
Fluid Mechanics and Its Applications
ISBN 978-3-032-28201-9 ISBN 978-3-032-28202-6 (eBook)
https://doi.org/10.1007/978-3-032-28202-6

This Springer imprint is published by the registered company Springer Nature Switzerland AG
The registered company address is: Gewerbestrasse 11, 6330 Cham, Switzerland

About the Book

This monograph presents a unified theoretical framework for predicting the shape and stability of pendent drops formed when a denser fluid overlies a lighter one in a gravitationally unstable configuration. The problem is formulated as a hydrostatic equilibrium governed by gravity and surface tension, with the interface pinned to a prescribed boundary. Central to the analysis is Rayleigh's work principle, whereby equilibrium shapes arise from the first variation of the potential energy, and stability is determined by the sign of the second variation.

Two complementary formulations are developed: one in which drop volume is prescribed and another in which pressure is controlled. In both cases, the governing Young–Laplace equation is derived and analyzed, and the solutions are characterized through volume–pressure (V–P) diagrams. These diagrams reveal the existence of critical points, identified as either turning points or branch points, corresponding respectively to symmetric and asymmetric modes of instability and breakup. A key result of the monograph is that these critical conditions—and hence stability limits—can be determined without explicitly solving eigenvalue problems, instead through geometric and analytical properties of the equilibrium solution curves.

The analysis begins with a one-dimensional model that captures the essential physics while enabling detailed analytical progress, including the classification of eigenmodes and the identification of instability mechanisms. This framework is then extended to pressure-controlled systems, where a direct correspondence between volume- and pressure-based descriptions is established. Finally, the theory is generalized to drops pinned to arbitrary planar curves, demonstrating how boundary geometry dictates symmetry and stability characteristics.

The work reveals a rich solution structure, including the emergence of branches of solutions and the interplay between geometry, control parameters, and instability. By showing that the stability problem can be resolved through the study of equilibrium shapes alone, this monograph provides both a conceptually elegant and practically powerful approach to understanding stability of fluid bodies in equilibrium.

Opening Remarks

The aim of this long essay is to present what we have learned about the shape and the stability of a pendent drop. To run a pendent drop experiment, a heavy fluid is made to lie above a light fluid. The fluids are in contact along a plane area bounded by a sharp-edged curve where the interface lying between the two fluids is pinned. This is the Rayleigh-Taylor problem. There is a critical Bond number such that the plane surface is stable to small displacements for all Bond numbers less than critical. This is also the zero volume pendent drop problem. To obtain a pendent drop we set the Bond number less than critical and then increase the volume of the heavy fluid, decrease the volume of the light fluid whereupon we have a pendent drop of a known volume. The inputs are the Bond number and the volume, the outputs are the drop shape and its stability. We plan to predict the drop shape and its stability given the Bond number and the drop volume.

We can also run a drop experiment by controlling the pressure acting on the heavy fluid. In this case the Bond numbers and the pressure are the inputs and the drop shape, its volume, and its stability are the outputs.

Pinning the drop to a sharp-edged curve tells us the symmetry of the drop in terms of the symmetry of the curve. This leads us to a one-dimensional model from which we can derive all that we wish to know about the stability of pendent drops.

The pendent drop is a study in stability of an equilibrium base state where the central idea is Rayleigh's work principle. Thus the first variation of the potential energy tells us about the drop shape, and the second variation tells us about its stability. We propose an eigenvalue problem such that perturbations that are multiples of an eigenfunction leads to eigenvalues, which tell us whether the second variation of the potential energy is positive or negative. This solves the stability problem. Critical points may be branch points or turning points. Upon, say, increasing the drop volume from zero we may find a branch point or a turning point before a branch point. This depends on the Bond number. We find that having solved the volume control problem a simple geometric construction solves the pressure control problem. In fact branch points under volume control predict branch points under pressure control.

The important ideas are equilibrium, symmetry and potential energy. This long essay ought to be viewed as the source of several lectures that can be presented to a class of graduate students.

Contents

1 The Shape and the Stability of a Pendent Drop Under Volume Control 1
1.1 Introduction to the Problem 1
1.2 A One-dimensional Model 3
1.3 The Potential Energy 4
1.4 Scaling 7
1.5 The Eigenvalue Problem 7
1.6 Plan 9
1.7 Do We Need to Solve an Eigenvalue Problem to Obtain Critical Points? 11
1.8 Is There Another Branch? 13
1.9 The New Solution Branch Near the Branch Point 15
1.10 Our Main Result 18
1.11 End Notes for Chap. 1 18
1.11.1 An Unconstrained Eigenvalue Problem 18
1.11.2 How Do Our V vs P Curves Start Out from $V = 0 = P$? 21
1.11.3 A Drop Slightly Misaligned 23
1.11.4 The Shape of Pendent Drops at Low Bond Numbers and Large Volumes 24

2 The Shape and the Stability of a Pendent Drop, Now Under Pressure Control 31
2.1 Introduction to the Problem 31
2.2 The Drop Shape Equation 32
2.3 Solving the Drop Shape Equation 35
2.4 Does the V vs P Curve Start Out in the First or the Second Quadrant? 37
2.5 Drawing the V *vs* P Curve at Input Values of B and B_* 38
2.6 Potential Energy 42
2.7 Critical Points 49

2.8 Branch Points, Turning Points 51
2.9 Locating Branch Points 53
2.10 Do Branch Points Come Before Turning Points? 55
2.11 Conclusion 58
2.12 End Notes for Chap. 2 59
2.12.1 Solving the Eigenvalue Problem: Pressure Control 59

3 The Stability of a Drop Pinned to an Arbitrary Plane Curve: Volume Control 65
3.1 Drop Pinned to a Circle 65
3.1.1 Introduction 65
3.1.2 A Few Formulas 66
3.1.3 The Equation for the Drop Shape 67
3.1.4 The Plan 68
3.1.5 The Eigenvalue Problem 69
3.1.6 Do We Need to Solve Eigenvalue Problems? 72
3.1.7 What Do the Two-dimensional Problem and the One-dimensional Model Have in Common? 74
3.2 Drops Pinned to a Closed Curve 75
3.3 A Drop Pinned to a Sharp-edged Rectangle 79
3.4 Conclusion 83

Chapter 1
The Shape and the Stability of a Pendent Drop Under Volume Control

Abstract This chapter develops a comprehensive framework for predicting the shape and stability of a pendent drop when the drop volume is the control parameter. Starting from the gravitationally unstable configuration of a heavy fluid over a lighter one, the drop is modeled as a hydrostatic equilibrium governed by the balance of gravity and surface tension. A one-dimensional model is introduced, enabling analytical tractability while preserving the essential physics. The governing Young–Laplace equation is derived via a variational formulation, where the first variation of potential energy yields the equilibrium shape and the second variation determines stability.

Stability is characterized through an eigenvalue problem, with the sign of eigenvalues indicating either stability or instability. The evolution of solutions is described in terms of volume-pressure (V–P) curves, where critical points arise as either turning points (associated with symmetric breakup) or branch points (associated with asymmetric breakup). A key result is that these critical points can be identified without explicitly solving the eigenvalue problem, using only the V–P curves. The chapter also reveals the existence of non-symmetric solution branches, emerging from branch points. The main conclusion is that the stability problem can be fully resolved through the analysis of equilibrium shapes alone.

1.1 Introduction to the Problem

We begin by presenting Fig. 1.1 where a heavy fluid lies above a light fluid. The fluids are immiscible. A drop of heavy fluid displaces light fluid. The drop is pinned to a sharp-edged circle. At the left, heavy fluid is added and light fluid is removed. Our aim is to obtain the shape of the drop, predict its stability and sketch the pattern at breakup.

We assume a drop shape is an equilibrium figure and so too every displacement of this shape, i.e., the drop shape problem is a hydrostatic problem. The forces acting are due to gravity and surface tension. Gravity creates pressure gradients in the two fluids. The pressure difference across the surface is balanced by forces acting normal

L. Johns, R. Narayanan, *The Shape and the Stability of Pendent Drops*,
Fluid Mechanics and Its Applications 144,
https://doi.org/10.1007/978-3-032-28202-6_1

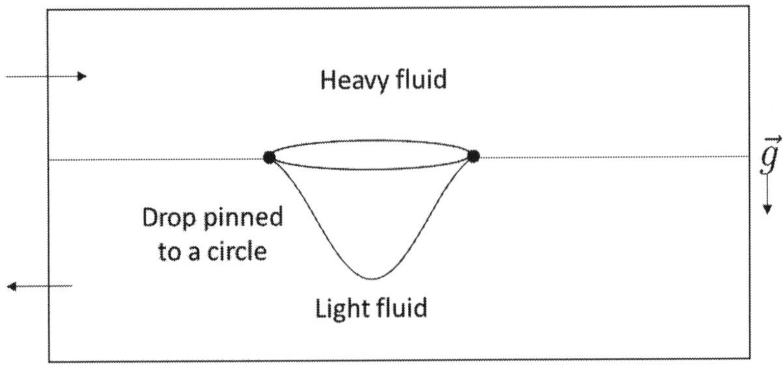

Fig. 1.1 Drop pinned to a circle

to the surface whose strength depends on the surface tension and the curvature of the surface. Gravity is destabilizing due to the adverse density difference. Surface tension is stabilizing. In balance they create an equilibrium drop shape. The drops grow from a plane surface separating the heavy and light fluids upon the addition of heavy fluid and the removal of light fluid. Deriving the stability of the plane surface is known as the Rayleigh-Taylor problem. Solving this problem reveals a critical diameter above which the heavy over light arrangement is unstable. Assuming the diameter is less than critical, this plane surface, the zero volume drop shape, is stable to small displacements. Our plan then is to add heavy fluid, remove light fluid, thus creating a drop of increasing volume. We thereby weaken the stability until we reach the limit of stability whereupon the drop is just about to break. The drop volume at this point is said to be the critical volume and our job is to predict this critical volume.

Now the input or the control variable may be the drop volume, as in Fig. 1.1, or it may be the pressure exerted on the heavy fluid. Thus, displacements of equilibrium drop shapes must be at constant volume or at constant pressure. We begin by working out the case where the input is the drop volume. We say that the drop is under volume control.

Associated to a drop shape is a potential energy. The potential energy of a drop does not change to first order in response to a small displacement. At second order, if it increases for all small displacements, the drop is stable. If it decreases for one small displacement, the drop is unstable. Thus the drop is at a critical point if we can find one small displacement such that, at second order, the potential energy does not change, though it increases for all other small displacements. This is Rayleigh's work principle. It solves the stability problem.

1.2 A One-dimensional Model

The drop pinned to a circle in Fig. 1.1 is a two-dimensional problem. We present the one-dimensional model sketched in Fig. 1.2 where the notation is introduced. This model tells us almost everything we might wish to know about a drop pinned to a sharp-edged circle.

The drop shape is denoted $z = Z(x)$ where there is no y variation. The drop volume then is per unit length in the y direction. The function Z is assumed to be a single-valued function of x and this is ordinarily the case. Thus we rule out cases where volumes are large and where drop shapes are nearly pieces of circles. The case where a vertical tangent obtains and $Z(x)$ is no longer single valued is addressed in end note 1.11.4.

The normal to the drop surface and the mean curvature of the surface depend on $Z(x)$ and they are given by the formulas:

$$\vec{n} = \frac{\vec{k} - Z_x \vec{i}}{\left(1 + Z_x^2\right)^{\frac{1}{2}}} \tag{1.1}$$

and

$$2H = \frac{d}{dx} \frac{Z_x}{\left(1 + Z_x^2\right)^{\frac{1}{2}}} \tag{1.2}$$

The drop volume, denoted V, is our input variable and our first job is to derive the equation for the drop shape, $Z(x)$, given the input value of V. The pressures are hydrostatic and they are given by

$$p = -\rho g z + \mathscr{C} \tag{1.3}$$

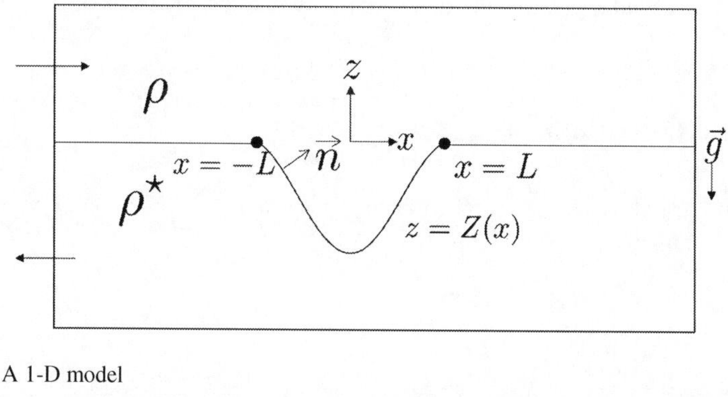

Fig. 1.2 A 1-D model

and

$$p^\star = -\rho^\star g z + \mathscr{C}^\star \tag{1.4}$$

Then the shape of the drop is the solution to the Young-Laplace equation, viz.,

$$p^\star + \gamma 2H = p \tag{1.5}$$

whereupon our equation for $Z(x)$ is

$$\mathscr{C} - \Delta\rho\, gZ = \gamma \frac{d}{dx} \frac{Z_x}{\left(1 + Z_x^2\right)^{\frac{1}{2}}} \tag{1.6}$$

where $\mathscr{C}$ denotes $\mathscr{C} - \mathscr{C}^\star$, where $Z = 0$ at $x = \pm L$, due to the drop being pinned at its points of contact, and where

$$V = -\int_{-L}^{L} Z(x) dx \tag{1.7}$$

In Eq. (1.6) $\Delta\rho$ denotes $\rho - \rho^\star > 0$ and $\mathscr{C}$ must be set so that the constraint, Eq. (1.7), is satisfied.

This problem has solutions, $Z(x)$, even in x. At first we direct our attention to these even solutions. Then we obtain a branch of non-even solutions, where having one such solution we have another due to our problem remaining unchanged if x is replaced by -x.

We turn our attention to the potential energy of a drop. The drop volume, denoted V_0, is set and is held fixed as the drop is given an equilibrium displacement, taken to be an input. Denoting its shape by $Z_0(x)$, where $Z_0 = 0$ at $x = \pm L$ and where $\int_{-L}^{L} Z_0(x)\, dx = -V_0$, we first derive Eq. (1.6) by assuming that $Z_0(x)$ is the shape that makes the potential energy stationary at first order in the displacement. Then by going to second order in the displacement we derive a formula for the second derivative of the potential energy at $Z_0(x)$ where we already know its first derivative is zero. The second derivative obtains given only the first order displacement.

1.3 The Potential Energy

Taking the gravitational potential to be gz, the gravitational potential energy of a pendent drop is given by

$$\int_{-L}^{L} dx \int_{Z}^{0} \rho g z dz = -\int_{-L}^{L} \frac{1}{2} \rho g Z^2 dx \tag{1.8}$$

whereupon, accounting for the buoyancy due to the light fluid and adding the surface potential energy, we have

$$PE(Z) = \int_{-L}^{L} \{\gamma(1 + Z_x{}^2)^{\frac{1}{2}} - \frac{1}{2}\Delta\rho g Z^2\} dx \tag{1.9}$$

Then denoting by $Z_0(x)$ the shape that makes $PE(Z)$ stationary at the volume, V_0, and denoting by $Z(x, \epsilon)$ a nearby equilibrium shape we write $Z(x, \epsilon) = Z_0(x) + \epsilon Z_1(x) + \frac{1}{2}\epsilon^2 Z_2(x)$ and denote $PE(Z)$ and $V(Z)$ by $PE(\epsilon)$ and $V(\epsilon)$.

Our first job is to obtain the equation satisfied by $Z_0(x)$. To do this we differentiate $PE(\epsilon)$ and $V(\epsilon)$ with respect to ϵ whereupon we have

$$\frac{dPE}{d\epsilon} = \int_{-L}^{L} \left(\gamma \frac{Z_x}{(1 + Z_x{}^2)^{\frac{1}{2}}} \frac{dZ_x}{d\epsilon} - \Delta\rho g Z \frac{dZ}{d\epsilon}\right) dx \tag{1.10}$$

$$\begin{aligned} \frac{d^2PE}{d\epsilon^2} = \int_{-L}^{L} \Bigg(\gamma \frac{1}{(1 + Z_x)^{\frac{3}{2}}} \frac{dZ_x}{d\epsilon}\frac{dZ_x}{d\epsilon} + \gamma \frac{Z_x}{(1 + Z_x{}^2)^{\frac{1}{2}}} \frac{d^2Z_x}{d\epsilon^2} \\ -\Delta\rho g \frac{dZ}{d\epsilon}\frac{dZ}{d\epsilon} - \Delta\rho g Z \frac{d^2Z}{d\epsilon^2}\Bigg) dx \end{aligned} \tag{1.11}$$

$$\frac{dV}{d\epsilon} = -\int_{-L}^{L} \frac{dZ}{d\epsilon} dx \tag{1.12}$$

and

$$\frac{d^2V}{d\epsilon^2} = -\int_{-L}^{L} \frac{d^2Z}{d\epsilon^2} dx \tag{1.13}$$

The constraints, due to holding the drop volume fixed, obtain upon setting both $\frac{dV}{d\epsilon}\big|_{\epsilon=0}$ and $\frac{d^2V}{d\epsilon^2}\big|_{\epsilon=0}$ to zero. Thus $Z_1(x)$ and $Z_2(x)$, where $Z_1 = \frac{dZ}{d\epsilon}\big|_{\epsilon=0}$ and $Z_2 = \frac{d^2Z}{d\epsilon^2}\big|_{\epsilon=0}$, must satisfy $\int_{-L}^{L} Z_1(x)\, dx = 0 = \int_{-L}^{L} Z_2(x)\, dx$

Now at $\epsilon = 0$ we have

$$\frac{dPE}{d\epsilon}\bigg|_{\epsilon=0} = \int_{-L}^{L} \left(\gamma \frac{Z_{0x}}{(1 + Z_{0x}{}^2)^{\frac{1}{2}}} Z_{1x} - \Delta\rho g Z_0 Z_1\right) dx \tag{1.14}$$

and integrating by parts and setting $Z_1(x = \pm L)$ to zero, we obtain

$$\frac{dPE}{d\epsilon}\bigg|_{\epsilon=0} = \int_{-L}^{L} \left(-\gamma \frac{d}{dx} \frac{Z_{0x}}{(1 + Z_{0x}{}^2)^{\frac{1}{2}}} - \Delta\rho g Z_0\right) Z_1 dx \tag{1.15}$$

Then in order that $PE(Z_0)$ be the same or less than $PE(Z)$ for all Z near Z_0 we must have

$$\left.\frac{dPE}{d\epsilon}\right|_{\epsilon=0} = 0 \tag{1.16}$$

no matter Z_1, so long as

$$\int_{-L}^{L} Z_1 dx = 0 \quad \text{and} \quad Z_1(x = \pm L) = 0 \tag{1.17}$$

Thus the factor in the integrand in Eq. (1.15) , viz.,

$$-\gamma \frac{d}{dx} \frac{Z_{0x}}{(1 + {Z_{0x}}^2)^{\frac{1}{2}}} - \Delta\rho g Z_0 \tag{1.18}$$

must be a constant, say $-\mathscr{C}_0$, and we obtain Eq. (1.6) for $Z_0(x)$ and $\mathscr{C}_0$ where

$$Z_0(x = \pm L) = 0 \tag{1.19}$$

and

$$\int_{-L}^{L} Z_0(x)dx = -V_0 \tag{1.20}$$

Now, given V_0, and having obtained $Z_0(x)$ where $\left.\frac{dPE}{d\epsilon}\right|_{\epsilon=0} = 0$ we wish to know if the potential energy increases, decreases or remains constant to second order in small equilibrium displacements. Our aim then is to obtain $\left.\frac{d^2PE}{d\epsilon^2}\right|_{\epsilon=0}$. To do this we turn to Eq. (1.11) and set ϵ to zero whereupon we have

$$\begin{aligned}\left.\frac{d^2PE}{d\epsilon^2}\right|_{\epsilon=0} &= \int_{-L}^{L} \left(\gamma \frac{Z_{1x}}{(1 + {Z_{0x}}^2)^{\frac{3}{2}}} Z_{1x} - \Delta\rho g Z_1 Z_1\right) dx \\ &+ \int_{-L}^{L} \left(\gamma \frac{Z_{0x}}{(1 + {Z_{0x}}^2)^{\frac{1}{2}}} Z_{2x} - \Delta\rho g Z_0 Z_2\right) dx\end{aligned} \tag{1.21}$$

Then, integrating by parts, setting $Z_1(\pm L)$ and $Z_2(\pm L)$ to zero, noticing that $Z_0(x)$ satisfies Eq. (1.6) and using $\int_{-L}^{L} Z_2(x)dx = 0$, we obtain

$$\left.\frac{d^2PE}{d\epsilon^2}\right|_{\epsilon=0} = \int_{-L}^{L} \left(-\gamma \frac{d}{dx} \frac{Z_{1x}}{(1 + {Z_{0x}}^2)^{\frac{3}{2}}} - \Delta\rho g Z_1\right) Z_1 dx \tag{1.22}$$

Then to answer the stability question we solve an eigenvalue problem to obtain a set of allowable displacements. But first we write our problems in scaled variables.

1.4 Scaling

To write our problem in scaled variables we scale Z and x by L and denote $\frac{\Delta\rho g L^2}{\gamma}$ and $\frac{\mathscr{C}L}{\gamma}$ by B and P. Then, at input values of B and V, to obtain the drop shape $Z(x)$ we must solve the problem

$$P - BZ = \frac{d}{dx}\frac{Z_x}{(1+Z_x^2)^{\frac{1}{2}}} \tag{1.23}$$

where $Z = 0$ at $x = \pm 1$ and where $\int_{-1}^{1} Z(x)dx = -V$. Then having the solution $Z(x)$ at B and V and assuming the shape $Z(x)$ is given a small displacement, $Z_1(x)$, we have, at $Z(x)$,

$$\left.\frac{d^2 PE}{d\epsilon^2}\right|_{\epsilon=0} = \int_{-1}^{1}\left\{-\frac{d}{dx}\frac{Z_{1x}}{(1+Z_{0x}^2)^{\frac{3}{2}}} - BZ_1\right\}Z_1 dx \tag{1.24}$$

1.5 The Eigenvalue Problem

To find out if a drop shape $Z(x)$ at a volume V is stable we plan to evaluate $\left.\frac{d^2 PE}{d\epsilon^2}\right|_{\epsilon=0}$ at a set of displacements solving an eigenvalue problem. It is

$$C - B\psi = \frac{d}{dx}\frac{\psi_x}{(1+Z_x^2)^{\frac{3}{2}}} + \lambda^2\psi \tag{1.25}$$

where $\psi = 0$ at $x = \pm 1$ and $\int_{-1}^{1}\psi dx = 0$ and where $Z(x)$ is known. The eigenvalues are denoted λ^2, the eigenfunctions are denoted $\psi(x)$, C. The value of C must be set so that the constraint $\int_{-1}^{1}\psi(x)dx = 0$ is satisfied. We can obtain C in terms of $\psi(x)$ via

$$2C = \left.\frac{\psi_x}{(1+Z_x^2)^{\frac{3}{2}}}\right|_{-1}^{1} \tag{1.26}$$

whereupon due to ${Z_x}^2(x = 1) = {Z_x}^2(x = -1)$ we have $C = 0$ if $\psi(x)$ is an odd function of x and $C \neq 0$ if $\psi(x)$ is an even function of x, assuming $\psi_x(x)$ is not zero at $x = \pm 1$.

For any displacement $Z_1(x)$ we have

$$\left.\frac{d^2 PE}{d\epsilon^2}\right|_{\epsilon=0} = -\int_{-1}^{1} \left(\frac{d}{dx} \frac{Z_{1x}}{(1 + Z_x^2)^{\frac{3}{2}}} + BZ_1 \right) Z_1 dx \tag{1.27}$$

and upon setting $Z_1(x)$ to an eigenfunction $\psi(x)$ where

$$\frac{d}{dx} \frac{\psi_x}{(1 + Z_x^2)^{\frac{3}{2}}} + B\psi = C - \lambda^2 \psi \tag{1.28}$$

we have

$$\boxed{\left.\frac{d^2 PE}{d\epsilon^2}\right|_{\epsilon=0} = \lambda^2 \int_{-1}^{1} \psi^2 dx} \tag{1.29}$$

whereupon $\left.\frac{d^2 PE}{d\epsilon^2}\right|_{\epsilon=0}$ is positive and our drop is stable if all the eigenvalues are positive. If one eigenvalue is zero, the others remaining positive, our drop is at a critical point. If at least one eigenvalue is negative our drop is unstable.

To have a stable drop at $V = 0$ we obtain the value of B such that all eigenvalues are positive there. To do this we solve the eigenvalue problem at $V = 0$ where we have $Z(x) = 0$. It is

$$C - B\psi = \frac{d^2\psi}{dx^2} + \lambda^2 \psi \tag{1.30}$$

where $\psi = 0$ at $x = \pm 1$ and where $\int_{-1}^{1} \psi(x) dx = 0$. Its odd solutions are

$$\psi = A \sin(\lambda^2 + B)^{\frac{1}{2}} x \tag{1.31}$$

Its even solutions are

$$\psi = A \left(\cos(\lambda^2 + B)^{\frac{1}{2}} x - \frac{\sin(\lambda^2 + B)^{\frac{1}{2}}}{(\lambda^2 + B)^{\frac{1}{2}}} \right) \tag{1.32}$$

The odd eigenvalues are

$$\lambda^2 = -B + \alpha^2, \quad \sin \alpha = 0 \tag{1.33}$$

The even eigenvalues are

$$\lambda^2 = -B + \alpha^2, \quad \tan\alpha = \alpha. \tag{1.34}$$

The least eigenvalue is

$$\lambda^2 = -B + \pi^2$$

and all eigenvalues are positive if B is less than π^2.

Due to $Z(x)$ being even in x, our eigenfunctions are either even or odd in x. The eigenvalues are called even or odd according as their eigenfunctions are even or odd. The order may even, odd, even $\cdots$ or odd, even, odd, $\cdots$. Both orders are possible depending on the values of B and V.

In end note 1.11.1 we solve our eigenvalue problem in terms of the solutions to an unconstrained eigenvalue problem, where the order of the eigenfunctions is always even, odd, even, $\cdots$.

1.6 Plan

Our plan is to set B to a value less than π^2 and to start at $V = 0$ where $Z(x) = 0$ and all our eigenvalues are positive. Then we advance V upwards from zero where at each V we obtain $Z(x)$, P and the least odd and the least even eigenvalues. We assume that as V advances our eigenvalues move to the left. We continue to advance V until the least eigenvalue reaches zero for the first time. This point is a critical point and the drop is at the point of breakup. The critical volume is denoted V_{crit}. The symmetry at breakup is due to whether the eigenfunction at $\lambda^2 = 0$ is even or odd.

At each V we have $Z(x)$ and P and we plot our results as a V vs P curve. In end note 1.11.2, we find that our curves start out in the first quadrant at $B < \frac{1}{4}\pi^2$, they start out in the second quadrant at $B > \frac{1}{4}\pi^2$. In Fig. 1.3 we present three V vs P curves. Along each curve V reaches its greatest value. This point where V is greatest is called a turning point.

Having arrived at a critical point we wish to know if we can continue to advance V, pass through V_{crit} and continue to solve the drop shape problem. We can if we can solve for $\dot{Z}$ where $\dot{Z} = \frac{dZ}{dV}$. The $\dot{Z}$ problem obtains on differentiating the drop shape problem with respect to V. It is

$$\dot{P} - B\dot{Z} = \frac{d}{dx}\frac{\dot{Z}_x}{(1 + Z_x^2)^{\frac{3}{2}}} \tag{1.35}$$

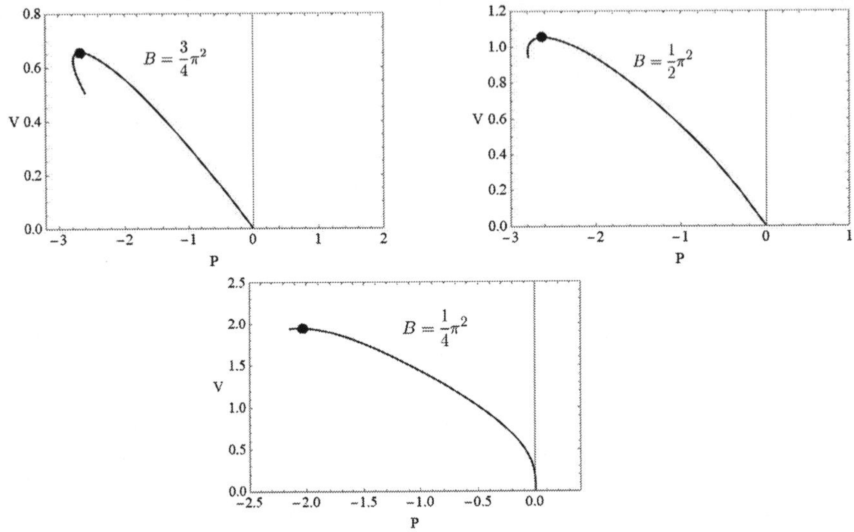

Fig. 1.3 Locating turning points for $B = \frac{3}{4}\pi^2$, $B = \frac{1}{2}\pi^2$ and $B = \frac{1}{4}\pi^2$

where $\dot{Z} = 0$ at $x = \pm 1$ and $\int_{-1}^{1} \dot{Z} dx = -1$. This is a linear, non-homogeneous problem, $\dot{Z}$ being driven by -1. At V_{crit}, the homogeneous $\dot{Z}$ problem has a non zero solution, viz., the eigenfunction corresponding to the eigenvalue $\lambda^2 = 0$. Before V reaches V_{crit} the homogeneous $\dot{Z}$ problem has only the solution zero and $\dot{Z}$ can be obtained. But at V_{crit} a solvability condition must be satisfied in order to obtain $\dot{Z}(x)$. To derive the solvability condition we multiply the equation for ψ at $\lambda^2 = 0$ by $\dot{Z}$, the equation for $\dot{Z}$ by ψ at $\lambda^2 = 0$, subtract and integrate over $(-1, 1)$. It is

$$C \int_{-1}^{1} \dot{Z} dx = -C = 0 \tag{1.36}$$

If the least odd eigenvalue is the first to reach zero, C is zero, solvability is satisfied and we can advance V passing through V_{crit}. The critical point is then a branch point. The drop is unsymmetric at breakup.

If the least even eigenvalue is the first to reach zero, C is not zero, solvability is not satisfied, we cannot obtain $\dot{Z}$ and we cannot advance V. Thus we have arrived at the greatest value of V and $V_{crit} = V_{greatest}$. The critical point is therefore a turning point. The drop is symmetric at breakup.

1.7 Do We Need to Solve an Eigenvalue Problem to Obtain Critical Points?

Our result thus far is this: to obtain critical points we need to solve an eigenvalue problem and find out where the least eigenvalue reaches zero. Our aim now is to obtain critical points by solving only the drop shape problem.

To obtain a turning point, given the value of B, we only need to solve for drop shapes and draw the V vs P curve. Then the turning point is the point where V is greatest. And we have found the point where the least even eigenvalue reaches zero without solving an eigenvalue problem. This eigenvalue is positive at $V = 0$. It remains positive as V advances upward reaching zero at the greatest value of V.

To obtain a branch point by solving only the drop shape problem, we begin by integrating the drop shape equation whereupon we have

$$2P + BV = \left. \frac{Z_x}{(1 + Z_x^2)^{\frac{1}{2}}} \right|_{-1}^{1} = 2\frac{Z_x}{(1 + Z_x^2)^{\frac{1}{2}}} \text{ at } x = 1 \tag{1.37}$$

This directs our attention to Z_x at $x = \pm 1$. In Fig. 1.4, we have marked the signs of the least odd and the least even eigenvalues along the V vs P curve at $B = \frac{3}{4}\pi^2$. We have also drawn the drop shapes at three points along the curve.

We see that where the least odd eigenvalue reaches zero, the drop has a horizontal tangent at the point of contact, i.e., $Z_x = 0$ at $x = \pm 1$, i.e., at the branch point the light fluid is just about to invade the heavy fluid.

To see that at branch points we always have horizontal tangents at points of contact we derive the equation for Z_x. To do this we differentiate the drop shape equation with respect to x whereupon we have

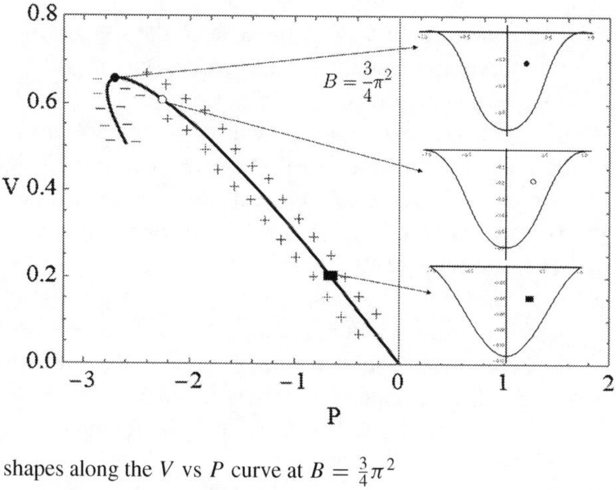

Fig. 1.4 Drop shapes along the V vs P curve at $B = \frac{3}{4}\pi^2$

$$-BZ_x = \frac{d}{dx}\frac{(Z_x)_x}{(1+Z_x^2)^{\frac{3}{2}}} \tag{1.38}$$

where $Z_x(-1) = -Z_x(1)$. Then, due to $C = 0$ and $\lambda^2 = 0$, we observe that the eigenfunction where the least odd eigenvalue reaches zero satisfies

$$-B\psi = \frac{d}{dx}\frac{\psi_x}{(1+Z_x^2)^{\frac{3}{2}}} \tag{1.39}$$

where $\psi = 0$ at $x = \pm 1$ and $\int_{-1}^{1}\psi dx = 0$. Thus from Eqs. (1.38) and (1.39) we have

$$\int_{-1}^{1}\left(\psi\frac{d}{dx}\frac{(Z_x)_x}{(1+Z_x)^{\frac{3}{2}}} - Z_x\frac{d}{dx}\frac{\psi_x}{(1+Z_x^2)^{\frac{3}{2}}}\right)dx = 0 \tag{1.40}$$

whereupon, by integration by parts, we obtain

$$Z_x\frac{\psi_x}{(1+Z_x^2)^{\frac{3}{2}}}\Bigg|_{-1}^{1} = 0 \tag{1.41}$$

Then, due to ψ_x being even and Z_x being odd, we find, at a branch point, that $Z_x = 0$ at $x = \pm 1$.

Now we do not plan to go looking for drops having horizontal tangents at their contact points. Instead from Eq. (1.37) we observe that at a branch point where $Z_x = 0$ at $x = \pm 1$ we have

$$2P + BV = 0 \tag{1.42}$$

and thus, setting B and drawing the V vs P curve at B, the straight line $2P + BV = 0$ crosses the V vs P curve at a branch point. Thus we only need to solve the drop shape problem to obtain branch points.

Our conclusion is this: we can obtain critical points by solving only the drop shape problem and then drawing the V vs P curve. The turning point, i.e., the greatest V along the V vs P curve is a critical point. The least even eigenvalue is zero there. The point where the line $2P + BV = 0$ crosses the V vs P curve is a branch point. The least odd eigenvalue is zero there.

Upon advancing V upward from zero, branch points may come before turning points or turning points may come before branch points. Which comes first depends on the value of B. The two cases are separated by the value of B, denoted $B_{\blacksquare}$, at which the least even and the least odd eigenvalues arrive at zero together. This value is slightly less than $\frac{1}{2}\pi^2$. In Fig. 1.5 we illustrate the identification of critical points.

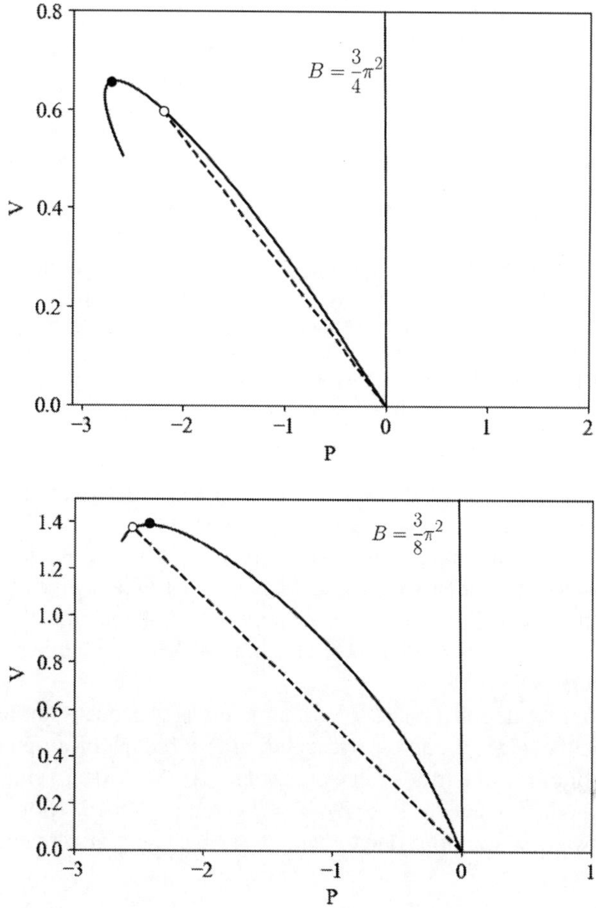

Fig. 1.5 Locating branch points and turning points for $B = \frac{3}{4}\pi^2$ and $B = \frac{3}{8}\pi^2$

1.8 Is There Another Branch?

Assuming that the critical volume is a branch point, we might expect another branch of solutions to pass through this point. The two branches we have, one coming in, the other going out, as V increases through the critical volume, correspond to even solutions to our drop shape problem. If there is yet another branch the straight line $V = -\frac{2}{B}P$ would be a likely candidate for its V vs P curve but we cannot imagine that the drop shapes will be even functions of x. To find out if we have a new class of solutions, we multiply Eq. (1.23) by Z_x and obtain

$$\frac{d}{dx}\left\{PZ - \frac{1}{2}BZ^2 + \frac{1}{(1 + Z_x^2)^{\frac{1}{2}}}\right\} = 0 \tag{1.43}$$

whereupon we have the first integral

$$PZ - \frac{1}{2}BZ^2 + \frac{1}{(1 + Z_x^2)^{\frac{1}{2}}} = constant \tag{1.44}$$

and we see that

$$Z_x^2(x = 1) = Z_x^2(x = -1) \tag{1.45}$$

Thus there are just two kinds of solutions. There are the even solutions where

$$Z_x(x = 1) = -Z_x(x = -1) \tag{1.46}$$

and there may be new solutions, not even, where

$$Z_x(x = 1) = Z_x(x = -1) \tag{1.47}$$

and if this is so their V vs P curve must be $2P + BV = 0$ which passes through our branch point. To find all possible solutions at a given value of B we can set P, set $Z = 0$ at $x = -1$ and integrate Eq. (1.23) to $x = 1$ for a family of values of Z_x at $x = -1$. We can then identify those Z_x's at $x = -1$ which wind up at $Z = 0$ at $x = 1$. Doing this for each P greater than but close to P at the critical point, we find three such Z_x's at $x = -1$. One leads to even drop shapes, the others lead to drop shapes, not even.

Thus the straight line that came to our attention as we sought out branch points turns out to be the V vs P curve corresponding to a branch of drop shapes which are not even solutions to our drop shape problem. At the critical value of V where $Z_x(\pm 1) = 0$ both $Z_x(x = 1) = \pm Z_x(x = -1)$ are satisfied. Along our straight line we have two new solutions due to our drop shape equation remaining the same if x is replaced by $-x$. We sketch non-even drop shapes in Fig. 1.6. We ought to

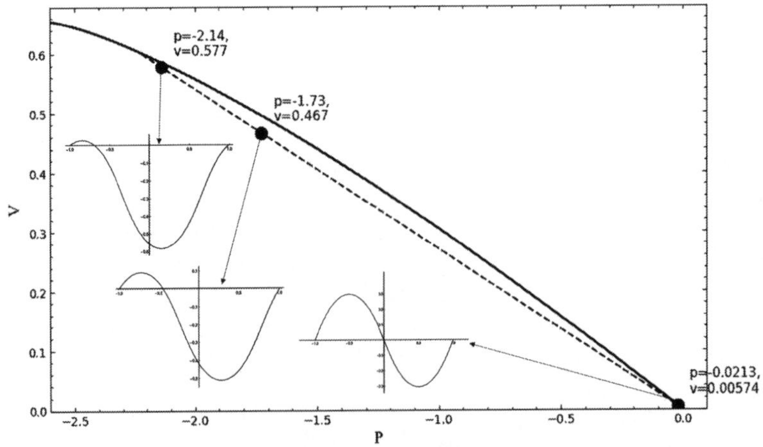

Fig. 1.6 Non-even drop shapes along the branch $2P + BV = 0$, $B = \frac{3}{4}\pi^2$

explain that the intersection of our two solution curves at the origin of the V vs P diagram in this figure is not a branch point. Here V is the input and P and $Z(x)$ are the outputs. At $V = 0$, we have $P = 0$ along both solution curves but the $Z(x)$'s are not the same. Thus, the $V, P, \|Z(x)\|$ curves do not cross at $V = 0$.

Now we might not have had the good luck to have discovered a first integral of our drop-shape equation but even so, we might have found the new solutions by expanding V about its critical value. We turn to this possibility.

1.9 The New Solution Branch Near the Branch Point

Having set B, we denote the even solution to the drop shape problem at the branch point V_0 by $Z_0(x)$ and P_0. The least eigenvalue is zero, it corresponds to an odd eigenfunction. We have $2P_0 + BV_0 = 0$ and $Z_{0x} = 0$ at $x = \pm 1$. To obtain a branch emerging from (P_0, V_0), other than the branches of even solutions we already have, we write $V = V_0 + \frac{1}{2}\epsilon^2 V_2$ where V_2 is an input moving V away from V_0. Then we expand Z and P as

$$Z(x, \epsilon) = Z_0(x) + \epsilon Z_1(x) + \frac{1}{2}\epsilon^2 Z_2(x) + \cdots \tag{1.48}$$

and

$$P(\epsilon) = P_0 + \epsilon P_1 + \frac{1}{2}\epsilon^2 P_2 + \cdots \tag{1.49}$$

and we write the formulas we will need, viz.,

$$\left.\frac{Z_x}{(1 + Z_x^2)^{\frac{1}{2}}}\right|_{\epsilon=0} = \frac{Z_{0x}}{(1 + Z_{0x}^2)^{\frac{1}{2}}} \tag{1.50}$$

$$\left.\frac{d}{d\epsilon}\frac{Z_x}{(1 + Z_x^2)^{\frac{1}{2}}}\right|_{\epsilon=0} = \frac{Z_{1x}}{(1 + Z_{0x}^2)^{\frac{3}{2}}} \tag{1.51}$$

$$\left.\frac{d^2}{d\epsilon^2}\frac{Z_x}{(1 + Z_x^2)^{\frac{1}{2}}}\right|_{\epsilon=0} = \frac{Z_{2x}}{(1 + Z_{0x}^2)^{\frac{3}{2}}} - 3\frac{Z_{0x}Z_{1x}^2}{(1 + Z_{0x}^2)^{\frac{5}{2}}} \tag{1.52}$$

and

$$\left.\frac{d^3}{d\epsilon^3}\frac{Z_x}{(1 + Z_x^2)^{\frac{1}{2}}}\right|_{\epsilon=0} = \frac{Z_{3x}}{(1 + Z_{0x}^2)^{\frac{3}{2}}} - \frac{9Z_{0x}Z_{1x}Z_{2x} + 3Z_{1x}^3}{(1 + Z_{0x}^2)^{\frac{5}{2}}} + 15\frac{Z_{0x}^2 Z_{1x}^3}{(1 + Z_{0x}^2)^{\frac{7}{2}}} \tag{1.53}$$

At the critical point we have $Z_0(x)$, the drop shape, and $\psi(x)$, the odd eigenfunction corresponding to $\lambda^2 = 0$. The first order problem is

$$P_1 - BZ_1 = \frac{d}{dx}\frac{Z_{1x}}{(1+Z_{0x}^2)^{\frac{3}{2}}} \tag{1.54}$$

where $Z_1 = 0$ at $x = \pm 1$ and $\int_{-1}^{1} Z_1 dx = -V_1 = 0$. Its solution is $Z_1 = A\psi$, $P_1 = 0$ where, therefore, $2P_1 + BV_1 = 0$. Our job is to obtain A. The second order problem is

$$P_2 - BZ_2 = \frac{d}{dx}\frac{Z_{2x}}{(1+Z_{0x}^2)^{\frac{3}{2}}} - 3\frac{d}{dx}\frac{Z_{0x}Z_{1x}^2}{(1+Z_{0x}^2)^{\frac{5}{2}}} \tag{1.55}$$

where $Z_2 = 0$ at $x = \pm 1$ and $\int_{-1}^{1} Z_2 dx = -V_2$, where V_2 is an input.

Because the homogeneous part of the Z_2 problem has the solution $Z_2 = \psi$, $P_2 = 0$, in order to obtain Z_2 a solvability condition must be satisfied. Writing $Z_1 = A\psi$, it is

$$\int_{-1}^{1} \psi \frac{d}{dx}\frac{Z_{0x}A^2\psi_x^2}{(1+Z_{0x}^2)^{\frac{5}{2}}}dx = -A^2\int_{-1}^{1}\frac{Z_{0x}\psi_x^3}{(1+Z_{0x}^2)^{\frac{5}{2}}}dx = 0 \tag{1.56}$$

and, due to Z_{0x} odd and ψ_x^3 even, it is satisfied for all A^2's whereupon A is not obtained at second order. To obtain a particular solution to the Z_2 problem we write

$$Z_2 = P_2 Z_2\Big|_{P_2=1,A^2=0} + A^2 Z_2\Big|_{P_2=0,A^2=1} \tag{1.57}$$

where $Z_2\Big|_{P_2=1,A^2=0}$ satisfies

$$1 = \frac{d}{dx}\frac{Z_{2x}}{(1+Z_{0x}^2)^{\frac{3}{2}}} + BZ_2 \tag{1.58}$$

where $Z_2\Big|_{P_2=0,A^2=1}$ satisfies

$$3\frac{d}{dx}\frac{Z_{0x}\psi_x^2}{(1+Z_{0x}^2)^{\frac{5}{2}}} = \frac{d}{dx}\frac{Z_{2x}}{(1+Z_{0x}^2)^{\frac{3}{2}}} + BZ_2 \tag{1.59}$$

and where $Z_2 = 0$ at $x = \pm 1$. We see that the particular solution to the Z_2 problem is an even function of x.

Now we have, where V_2 is an input,

$$-V_2 = \int_{-1}^{1} Z_2 dx = P_2 \int_{-1}^{1} Z_2\Big|_{P_2=1, A^2=0} dx + A^2 \int_{-1}^{1} Z_2\Big|_{P_2=0, A^2=1} dx \quad (1.60)$$

which gives us one equation in P_2 and A^2. By going to third order we expect to obtain a second equation in P_2 and A^2.

At third order our problem is

$$P_3 - BZ_3 = \frac{d}{dx}\frac{Z_{3x}}{(1+Z_{0x}^2)^{\frac{3}{2}}} - \frac{d}{dx}\{\frac{9Z_{0x}Z_{1x}Z_{2x} + 3Z_{1x}^3}{(1+Z_{0x}^2)^{\frac{5}{2}}} + 15\frac{Z_{0x}^2 Z_{1x}^3}{(1+Z_{0x}^2)^{\frac{7}{2}}}\} \quad (1.61)$$

and $Z_3 = 0$ at $x = \pm 1$. Again, a solvability condition must be satisfied. Thus in order that Z_3 can be obtained, we must have

$$0 = \int_{-1}^{1} \psi \frac{d}{dx}\{-9\frac{Z_{0x}Z_{1x}Z_{2x}}{(1+Z_{0x}^2)^{\frac{5}{2}}} - 3\frac{Z_{1x}^3}{(1+Z_{0x}^2)^{\frac{5}{2}}} + 15\frac{Z_{0x}^2 Z_{1x}^3}{(1+Z_{0x}^2)^{\frac{7}{2}}}\}dx \quad (1.62)$$

Integrating by parts, using $Z_1 = A\psi$ and $Z_2 = P_2 Z_2\Big|_{P_2=1, A^2=0} + A^2 Z_2\Big|_{P_2=0, A^2=1}$, we have

$$0 = P_2 A \int_{-1}^{1} \{-9\frac{Z_{0x}\psi_x^2 Z_{2x}\Big|_{P_2=1, A^2=0}}{(1+Z_{0x}^2)^{\frac{5}{2}}}\}dx + A^3 \int_{-1}^{1} \{-9\frac{Z_{0x}\psi_x^2 Z_{2x}\Big|_{P_2=0, A^2=1}}{(1+Z_{0x}^2)^{\frac{5}{2}}}$$
$$- 3\frac{\psi_x^4}{(1+Z_{0x}^2)^{\frac{5}{2}}} + 15\frac{Z_{0x}^2 \psi_x^4}{(1+Z_{0x}^2)^{\frac{7}{2}}}\}dx \quad (1.63)$$

where all the functions to be integrated are even in x. Evaluating the integrals we have our second equation in P_2 and A^2. Then assuming V_2 is negative we expect to obtain P_2 and a positive value of A^2 and we expect to learn that $2P_2 + BV_2$ is zero. Our hopes materialize and at $B = \frac{3}{4}\pi^2$ we have the results presented in Table 1.1. Due to $A = \pm\sqrt{A^2}$ we have two solutions.

Table 1.1 P_2 and A^2 at $V_2 = -1$ and $B = \frac{3}{4}\pi^2$

P_2	≈ 3.7
$2P_2 + BV_2$	≈ 0
A^2	≈ 584

1.10 Our Main Result

Our main result is this: we can solve the problem of the stability of one-dimensional drops by solving for no more than drop shapes. We never need to solve an eigenvalue problem. This result stems from three facts: the problem is hydrostatic, the drop is pinned and there is symmetry. We expect no less if the drop is pinned to a circle or if we input an outside pressure instead of the drop volume. We now direct our attention to these new cases.

1.11 End Notes for Chap. 1

1.11.1 An Unconstrained Eigenvalue Problem

Setting the values of B and V we have $Z(x)$ and P and our aim is to obtain the least eigenvalue. We wish to know if it is even or odd. Both cases are possible due to the fact that a constraint must be satisfied. To solve the eigenvalue problem we derive a formula for the eigenvalues, λ^2, in terms of the eigenvalues, denoted μ^2, which are solutions to an unconstrained eigenvalue problem where the least unconstrained eigenvalue is always even, the next larger is always odd, etc. It is

$$-B\phi = \frac{d}{dx}\frac{\phi_x}{(1+Z_x^2)^{\frac{1}{2}}} + \mu^2\phi \tag{1.64}$$

where $\phi = 0$ at $x = \pm 1$. We denote its solutions:

$$\begin{aligned} &\mu_1^2 < \mu_2^2 < \mu_3^2 < \ldots \\ &\phi_1\ ,\quad \phi_2\ ,\quad \phi_3, \ldots \\ &even\ , odd\ , even, \ldots \end{aligned}$$

At $V = 0$, where $Z(x)$ is zero, we have,

$$\begin{aligned} &\mu_1^2 = -B + \frac{1}{4}\pi^2 \\ &\mu_2^2 = -B + \pi^2 \\ &etc. \end{aligned}$$

Given V, all μ^2's and therefore all λ^2's are greater than $-B$. To see that this is so, multiply Eq. (1.64) by ϕ and integrate over $x \epsilon (-1, 1)$. Then integrating by parts and observing that $\phi = 0$ at $x = \pm 1$, we have

$$\mu^2 = -B + \frac{\int_{-1}^{1} \frac{\phi_x^2}{(1+Z_x^2)^{\frac{3}{2}}}}{\int_{-1}^{1} \phi^2 dx} \tag{1.65}$$

First we observe that the odd λ^2's are the odd μ^2's. Then we derive the even λ^2's from the even μ^2's. To do this we assume that

$$\int_{-1}^{1} \phi^2 dx = 1 \tag{1.66}$$

and expand any even ψ in the even ϕ's, viz,

$$\psi = \sum_{i=1,3,5,\cdots} c_i \phi_i \tag{1.67}$$

where

$$c_i = \int_{-1}^{1} \phi_i \psi dx \tag{1.68}$$

Then, multiplying Eq. (1.25) by ϕ_i, Eq. (1.64) by ψ, subtracting and integrating over (-1,1) we have

$$CI_i = (\lambda^2 - \mu_i^2)c_i \quad , i.e., \quad C\frac{I_i}{\lambda^2 - \mu_i^2} = c_i \tag{1.69}$$

where

$$I_i = \int_{-1}^{1} \phi_i dx \tag{1.70}$$

Upon multiplying Eq. (1.69) by ϕ_i, summing, integrating over (-1,1) and taking account of $\int_{-1}^{1} \psi dx = 0$ and $C \neq 0$, we have

$$LHS \equiv \sum_{i=1,3,\cdots} \frac{I_i^2}{(\lambda^2 - \mu_i^2)} = 0 \tag{1.71}$$

Now, given B and V, and therefore $Z(x)$, we have

$$\mu_1^2,\ \mu_3^2,\ \mu_5^2, \cdots$$
$$I_1^2,\ I_3^2\ ,\ I_5^2, \cdots$$

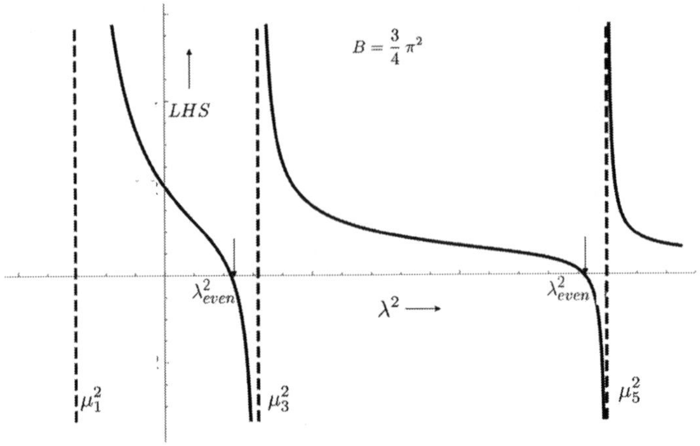

Fig. 1.7 A plot of LHS vs λ^2 giving the even λ^2's

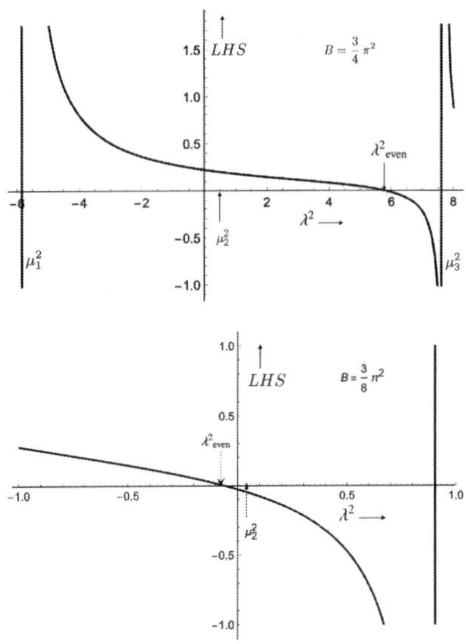

Fig. 1.8 The least even eigenvalue lying to the right of the least odd eigenvalue (top figure) and lying to the left of the least odd eigenvalue (bottom figure)

and by plotting $\sum_{i=1,3,\cdots} \frac{I_i^2}{\lambda^2-\mu_i^2}$ vs λ^2 we obtain the even λ^2's, cf., Fig. 1.7, where the even eigenvalues are located at the down-pointing arrows. They lie on the intervals (μ_1^2, μ_3^2),(μ_3^2, μ_5^2), $\cdots$

Now the least odd eigenvalue is μ_2^2 where $\mu_1^2 < \mu_2^2 < \mu_3^2$, and depending on the values of B and V the least even eigenvalue may lie to the left or to the right of μ_2^2, cf., Fig. 1.8.

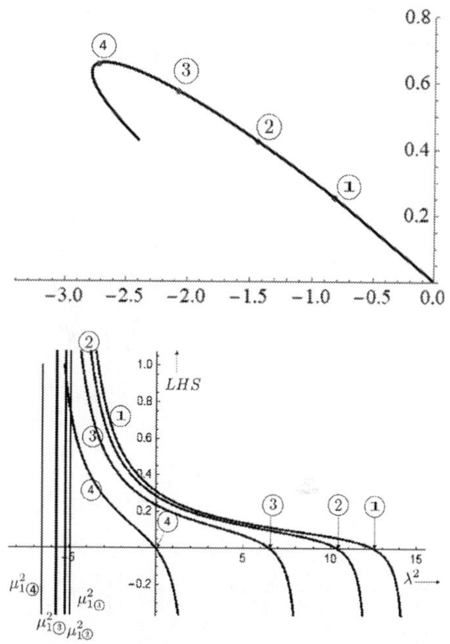

Fig. 1.9 The least even eigenvalue is positive until V reaches its greatest value where it is zero. The arrows point to the least even eigenvalue

We sketch the left-most curve of LHS vs λ^2 at four values of V at $B = \frac{3}{4}\pi^2$, cf. Fig. 1.9. The least even eigenvalue starts out positive at $V = 0$. It remains positive but decreases, as V increases until it reaches zero at the greatest value of V.

1.11.2 How Do Our *V* vs *P* Curves Start Out from *V* = 0 = *P* ?

To see how our V vs P curves start out from $P = 0$, $V = 0$ we turn to the equation for $\dot{Z}$. At $V = 0$, where the solution to our drop shape problem is $Z(x) = 0$, we have

$$\dot{P} - B\dot{Z} = \frac{d^2}{dx^2}\dot{Z} \tag{1.72}$$

and $\dot{Z} = 0$ at $x = \pm 1$ where $\int_{-1}^{1} \dot{Z} dx = -1$. Thus, assuming $\dot{Z}(x)$ is an even function of x, we have

$$\dot{Z}(x) = A \cos \sqrt{B} x + \frac{\dot{P}}{B} \tag{1.73}$$

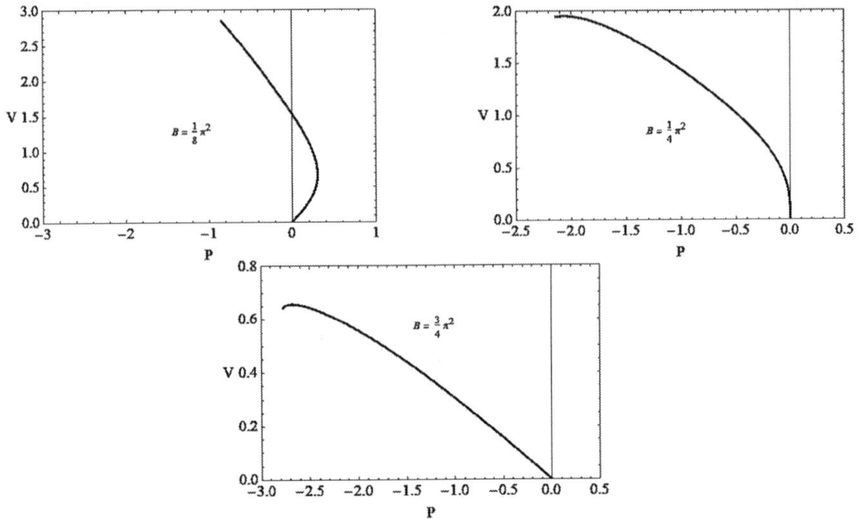

Fig. 1.10 V vs P curves start out in the first or second quadrants as B is less or greater than $\frac{1}{4}\pi^2$

and due to $\dot{Z} = 0$ at $x = \pm 1$,

$$\frac{\dot{P}}{B} = -A \cos \sqrt{B} \tag{1.74}$$

Then, due to $\int_{-1}^{1} \dot{Z} dx = -1$, we have

$$A = -\frac{1}{2} \frac{1}{\frac{\sin \sqrt{B}}{\sqrt{B}} - \cos \sqrt{B}} \tag{1.75}$$

and

$$\dot{P} = \frac{1}{2} B \frac{1}{\frac{\tan \sqrt{B}}{\sqrt{B}} - 1} \tag{1.76}$$

As B approaches zero, $\dot{P}$ approaches $\frac{3}{2}$. Thus for small values of B the V vs P curve starts out in the first quadrant. At $B = \frac{1}{4}\pi^2$ we have $\dot{P} = 0$ and the V vs P curve starts out along the V axis. For all B greater than $\frac{1}{4}\pi^2$ the curve starts out in the second quadrant. This is what we see in Fig. 1.10.

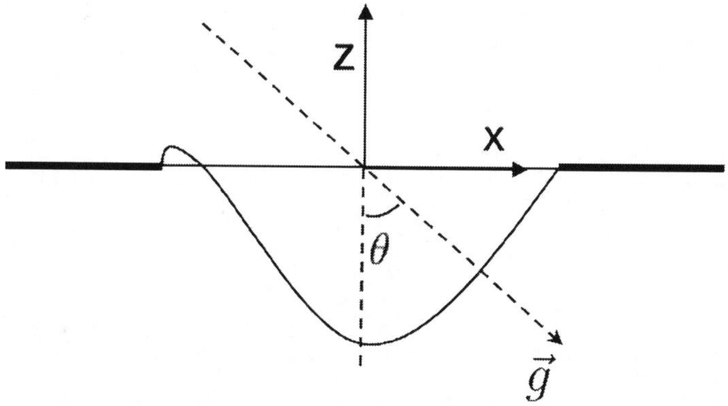

Fig. 1.11 A drop not perfectly aligned with gravity

1.11.3 A Drop Slightly Misaligned

In Fig. 1.11 we have a drop not perfectly aligned with gravity. It is slightly misaligned.

In scaled variables, the drop shape is obtained by solving

$$P + B\theta x - BZ = \frac{d}{dx}\frac{Z_x}{(1+Z_x^2)^{\frac{1}{2}}} \tag{1.77}$$

where $Z = 0$ at $x = \pm 1$ and

$$V = -\int_{-1}^{1} Z dx \tag{1.78}$$

and where we have replaced $cos\theta$ by 1 and $sin\theta$ by θ and where we notice that Z is no longer an even function of x.

We present a sketch in Fig. 1.12 of the V vs P curve at $B = \frac{1}{2}\pi^2$, assuming $\theta = \frac{1}{50}\pi$, where if the drop were perfectly aligned there would be a branch point just before the turning point.

The ψ and $\dot{Z}$ problems remain as they were, though Z is no longer an even function of x. We wish to have all eigenvalues positive at $V = 0$ and for small values of θ we imagine that the allowable B's are nearly as large as π^2. Upon advancing V upward from zero we expect to reach a value of V where the least λ^2 falls to zero for the first time. This V is critical and we wish to know if we can continue to advance V. We can if we can obtain $\dot{Z}$.

At $\lambda^2 = 0$ the homogeneous $\dot{Z}$ problem has the non-zero solution ψ. Thus a solvability condition must be satisfied. It is

$$C = 0 \tag{1.79}$$

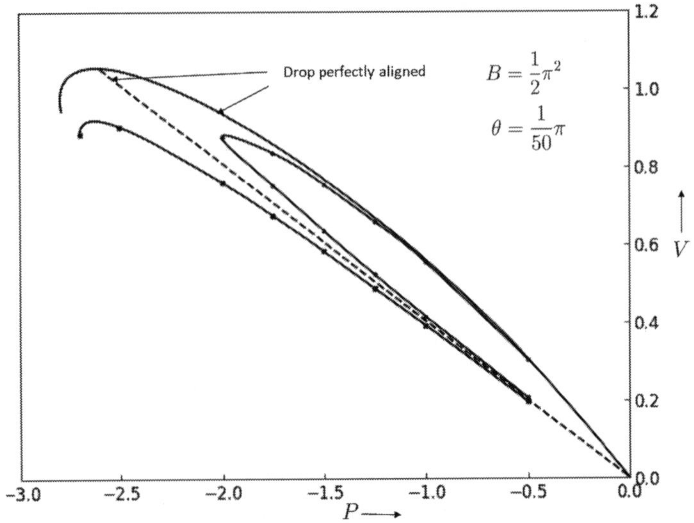

Fig. 1.12 V vs P curves at $\theta = 0$ and $\theta = \frac{1}{50}\pi$ at $B = \frac{1}{2}\pi^2$

To obtain C we integrate the eigenvalue problem obtaining

$$2C = \left. \frac{\psi_x}{(1 + Z_x^2)^{\frac{3}{2}}} \right|_{-1}^{1} \tag{1.80}$$

where if θ is not zero Z_x^2 is not the same at $x = \pm 1$ and we conclude that C is ordinarily not zero. Thus we cannot obtain $\dot{Z}$ and all critical points are turning points. For drops where a branch point appears before a turning point, if the drop is perfectly aligned, slightly slanting the drop leads to a turning point before the original branch point. In an experiment, where the drop is perfectly aligned, at least to our best effort, we expect to see drops breaking in what appears to be a non-symmetric pattern, much as if the critical point were a branch point.

1.11.4 The Shape of Pendent Drops at Low Bond Numbers and Large Volumes

Ordinarily, to find the shape of a pendent drop pinned to a sharp edge in the case of our one-dimensional model, viz., $z = Z(x)$, we solve

$$P - BZ = \frac{d}{dx} \frac{Z_x}{(1 + Z_x^2)^{\frac{1}{2}}} \tag{1.81}$$

$$Z = 0 \;\; at \;\; x = \pm 1 \tag{1.82}$$

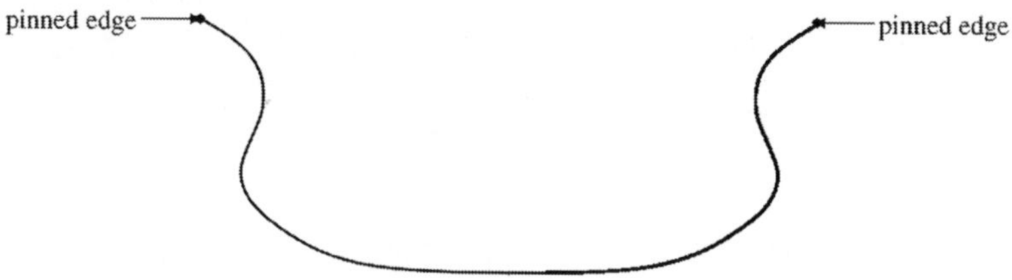

Fig. 1.13 At low Bond numbers a vertical tangent arises as V increases

and

$$V = -\int_{-1}^{1} Z(x)dx \tag{1.83}$$

The problem is written in scaled variables, the inputs are B and V and the outputs are P and $Z(x)$. No technical difficulties present themselves at small volumes. But if surface tensions are large and density differences are small, i.e., if Bond numbers are small, drop shapes approximate pieces of circles, and as we increase the volume of the drop we may find a vertical tangent, at a point where Z_x becomes infinitely large, cf. Fig. 1.13.

Our job is to find the shape of a drop under conditions where a vertical tangent is a likely prospect. To do this we cut the drop with a line, $z = z^*$, dividing it into two pieces, one where we write

$$x = X(z), z^* < z < 0 \tag{1.84}$$

and the other where we write

$$z = Z(x), 0 < x < X(z^*) \tag{1.85}$$

The two pieces are in contact at $x = X(z^*)$, $z = z^*$, where z^* is set such that Z_x is bounded along $z = Z(x)$, cf. Fig. 1.14.

Along $x = X(z)$ the normal is outward, along $z = Z(x)$ it is inward. The normals are given by

$$\frac{\vec{i} - X_z \vec{k}}{(1 + X_z^2)^{\frac{1}{2}}} \tag{1.86}$$

and

$$\frac{\vec{k} - Z_x \vec{i}}{(1 + Z_x^2)^{\frac{1}{2}}} \tag{1.87}$$

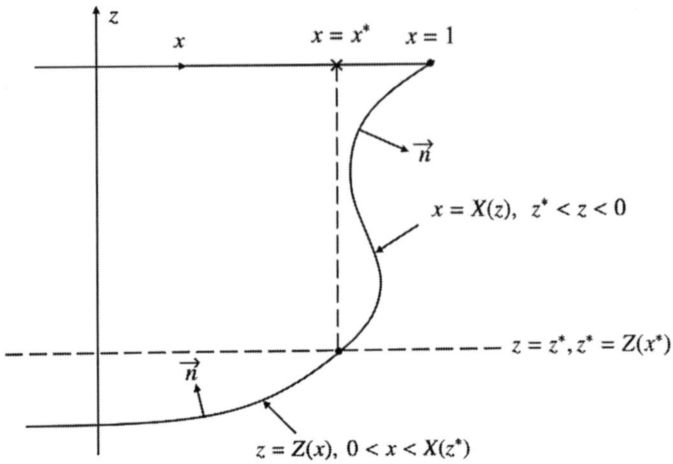

Fig. 1.14 Setting z^* and defining the shape via $z = Z(x)$, $z < z^*$ and $x = X(z)$, $z > z^*$

and to align the normals at the point of contact we must have

$$\frac{X_z}{(1+X_z^2)^{\frac{1}{2}}} = \frac{1}{(1+Z_x^2)^{\frac{1}{2}}} \tag{1.88}$$

and

$$\frac{1}{(1+X_z^2)^{\frac{1}{2}}} = \frac{Z_x}{(1+Z_x^2)^{\frac{1}{2}}} \tag{1.89}$$

whereupon we must have

$$Z_x = \frac{1}{X_z} \tag{1.90}$$

We assume left-right symmetry, direct our attention to the right half of the drop, and account only for the contact point at

$$x = x^* \;=\; X(z^*) > 0 \text{ and } z = z^* = Z(x^*) \tag{1.91}$$

Assuming the pressures are hydrostatic, the Young-Laplace equation gives us the equation for $Z(x)$. It is

$$P - BZ = \frac{d}{dx}\frac{Z_x}{(1+Z_x^2)^{\frac{1}{2}}}, \quad 0 < x < X(z^*) \tag{1.92}$$

where we have

$$Z_x = 0 \text{ at } x = 0 \tag{1.93}$$

Likewise we obtain the equation for $X(z)$. It is, again assuming that the pressures are hydrostatic and using the Young-Laplace equation,

$$P - Bz = -\frac{d}{dz}\frac{X_z}{(1 + X_z^2)^{\frac{1}{2}}}, \quad z^* < z < 0 \tag{1.94}$$

where we have $X = 1$ at $z = 0$.

Thus we have an equation for $X(z)$ and an end condition at $z = 0$ and we have an equation for $Z(x)$ and an end condition at $x = 0$. Where our solutions join, i.e., at $x = X(z^*),\ z = z^*$ we have

$$Z\big(X(z^*)\big) = z^* \tag{1.95}$$

and

$$Z_x(X(z^*))X_z(z^*) = 1 \tag{1.96}$$

The volume of the right hand half of the drop is given by

$$V = \int_{z^*}^{0} X(z)dz + \int_0^{X(z^*)} (z^* - Z(x))dx \tag{1.97}$$

To solve for the drop shape we set z^* and P. Then we guess a value for $X_z(z = 0)$ and solve for $X(z)$ on $z^* < z < 0$ obtaining $X(z^*)$ and $X_z(z^*)$. We then set $Z(X(z^*))$ to z^* and $Z_x(X(z^*))$ to $(1/X_z(z^*))$ and solve for $Z(x)$ on $0 < x < X(z^*)$. The check on the guess is $Z_x = 0$ at $x = 0$. Assuming the calculation converges we have the drop shape and its volume, given z^*.

Now the drop shape is given by

$$x = X(z), z^* < z < 0 \tag{1.98}$$

and

$$z = Z(x), 0 < x < X(z^*) \tag{1.99}$$

This drop shape must be independent of the choice of z^*, even though its representation depends on z^*.

To establish that the geometry of our problem is independent of z^*, we begin by looking at a curve in the $x - z$ plane, cf., Fig. 1.15.

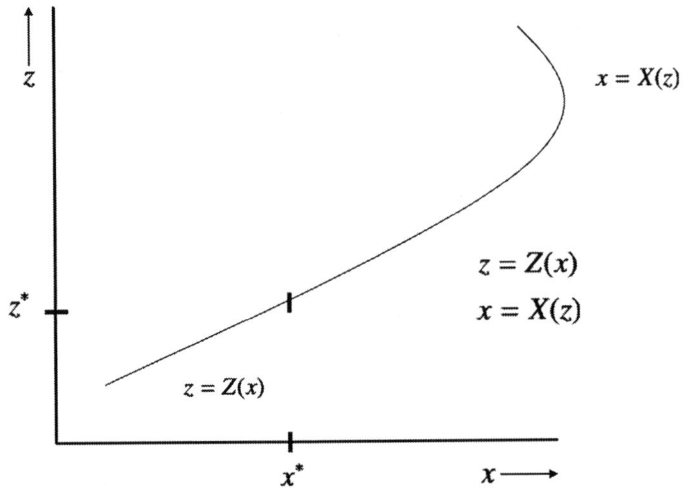

Fig. 1.15 Exchanging independent and dependent variables as we move along a curve

The points of the curve are denoted $(x, Z(x))$ where x is the independent variable and z is the dependent variable. Now along the curve, say at $x = x^*$, we exchange dependent and independent variables so that beyond $x = x^*$ we have two formulas for our curve, viz.,

$$z = Z(x) \tag{1.100}$$

and

$$x = X(z) \tag{1.101}$$

whereupon we have

$$dz = Z_x(x)dx \tag{1.102}$$

and

$$dx = X_z(z)dz \tag{1.103}$$

and thus

$$Z_x(x)X_z(z) = 1 \tag{1.104}$$

and

$$\frac{1}{1+{Z_x}^2} = \frac{X_z^2}{1+X_z^2} \tag{1.105}$$

We set x^* so that Z_x and X_z are positive in the neighbourhood of x^*.

Now the formula for our curve as it passes through $x = x^*$ is obtained by solving

$$P - BZ = \frac{d}{dx}\frac{Z_x}{(1+Z_x^2)^{\frac{1}{2}}} \tag{1.106}$$

where the right hand side is

$$\frac{Z_{xx}}{(1+Z_x^2)^{\frac{3}{2}}} \tag{1.107}$$

and we observe that

$$Z_{xx} = -\frac{1}{X_z^2}X_{zz}Z_x = -\frac{1}{X_z^3}X_{zz} \tag{1.108}$$

whereupon setting $Z(x)$ to z and using

$$\frac{1}{(1+{Z_x}^2)^{\frac{3}{2}}} = \frac{X_z^3}{(1+X_z^2)^{\frac{3}{2}}} \tag{1.109}$$

we obtain the equation for $X(z)$. It is

$$P - Bz = -\frac{d}{dz}\frac{X_z}{(1+X_z^2)^{\frac{1}{2}}} \tag{1.110}$$

Thus our two formulas

$$z = Z(x) \;\text{ and }\; x = X(z) \tag{1.111}$$

obtained by solving

$$P - BZ = \frac{d}{dx}\frac{Z_x}{(1+Z_x^2)^{\frac{1}{2}}} \tag{1.112}$$

and

$$P - Bz = -\frac{d}{dz}\frac{X_z}{(1+X_z^2)^{\frac{1}{2}}} \tag{1.113}$$

produce one and the same curve no matter the value of x^* we choose in order to introduce a second representation.

Chapter 2
The Shape and the Stability of a Pendent Drop, Now Under Pressure Control

Abstract This chapter extends the analysis of pendent drops to the case where pressure, rather than volume, is the control parameter. The drop shape is obtained by solving the Young–Laplace equation under prescribed pressure differences, with the resulting shape, volume, and stability emerging as outputs. The formulation emphasizes the construction of V–P curves for given Bond numbers, enabling a direct comparison with the volume-controlled case.

The chapter investigates the nature of critical points, branch points, and turning points, and provides methods for locating them systematically. A central question addressed is the ordering of these points along the V–P curve and how this depends on system parameters such as the Bond number. The analysis demonstrates that branch points under volume control correspond directly to branch points under pressure control, establishing a deep connection between the two formulations.

Energy methods and eigenvalue analysis are again used to determine stability, but the chapter highlights that much of the structure can be inferred geometrically from the V–P relationship. The results unify the two control paradigms and show that pressure-controlled experiments can be interpreted through the insights gained from volume control, providing a powerful and general framework for understanding pendent drop behavior.

2.1 Introduction to the Problem

The ground rules remain as they were. Our model is one-dimensional and hydrostatic conditions prevail. The volume of our drop is no longer the input variable, it is now a pressure. To explain the changes we must make we begin with a sketch of a drop under volume control, cf., Fig. 2.1.

We see that the drop has a well-defined volume. It is also the seat of a vertical pressure variation. But we cannot imagine controlling any of these pressures, where, to control a pressure, we must be able to set its value and then hold that value fixed if the drop is displaced from its equilibrium shape. Thus we need a new sketch and new drop shape equations. This will be our first job. We turn to the case where an

L. Johns, R. Narayanan, *The Shape and the Stability of Pendent Drops*,
Fluid Mechanics and Its Applications 144,
https://doi.org/10.1007/978-3-032-28202-6_2

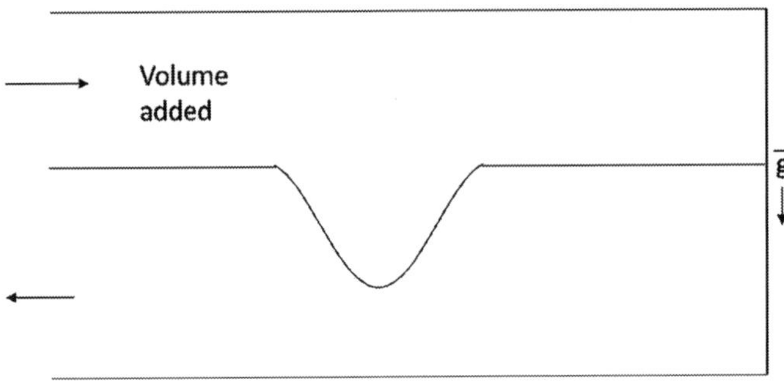

Fig. 2.1 A drop under volume control

external pressure is the input variable and the drop shape and its volume are the output variables.

2.2 The Drop Shape Equation

In Fig. 2.2 we present a sketch of a one-dimensional drop, no y variation, whose shape and volume are responding to an input pressure, denoted p, where p_{ref} is a fixed reference pressure. The densities of the heavy and the light fluids are denoted ρ and $\rho^{\star}$. All volumes are per unit depth.

To obtain the shape of the drop, viz., $z = Z(x)$, we must solve a hydrostatic problem. The pressures in the fluids are

$$p + \rho g(h - z) \tag{2.1}$$

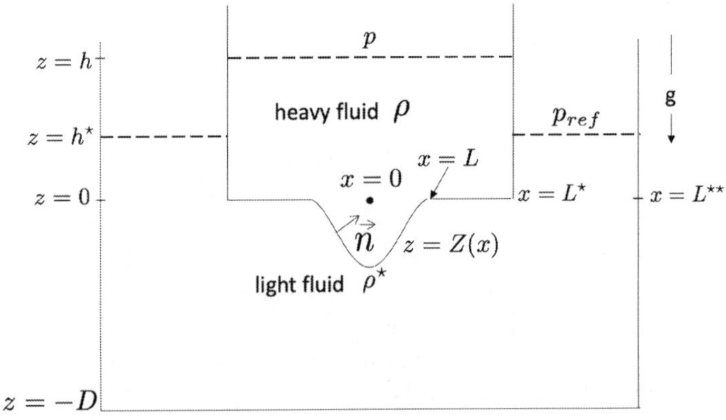

Fig. 2.2 A drop under pressure control

and

$$p_{ref} + \rho^{\star} g(h^{\star} - z) \tag{2.2}$$

We denote the volumes of the heavy and the light fluids by V_{heavy} and V_{light}. They are fixed and are given in terms of V, the drop volume, by

$$V_{heavy} = 2L^{\star}h + V \tag{2.3}$$

and

$$V_{light} = 2L^{\star\star}(h^{\star} + D) - 2L^{\star}h^{\star} - V \tag{2.4}$$

where V_{heavy}, V_{light}, L, $L^{\star}$, $L^{\star\star}$ and p_{ref} are inputs, part of the design of an experiment. The normal to the drop surface and its mean curvature are given by

$$\vec{n} = \frac{\vec{k} - Z_x \vec{i}}{(1 + Z_x^2)^{\frac{1}{2}}} \tag{2.5}$$

and

$$2H = \frac{d}{dx} \frac{Z_x}{(1 + Z_x^2)^{\frac{1}{2}}} \tag{2.6}$$

The drop shape is then the solution to the equation

$$p_{ref} + \rho^{\star} g(h^{\star} - Z) + \gamma 2H = p + \rho g(h - Z) \tag{2.7}$$

and we write this

$$p - p_{ref} + \rho g h - \rho^{\star} g h^{\star} - (\rho - \rho^{\star}) g Z = \gamma \frac{d}{dx} \frac{Z_x}{(1 + Z_x^2)^{\frac{1}{2}}} \tag{2.8}$$

where, having V_{heavy} and V_{light}, we can eliminate h and $h^{\star}$ in favor of V, viz.,

$$h = \frac{V_{heavy} - V}{2L^{\star}} \tag{2.9}$$

and

$$h^{\star} = \frac{V_{light} - 2L^{\star\star}D + V}{2(L^{\star\star} - L^{\star})} \tag{2.10}$$

whereupon we have

$$p - p_{ref} + \frac{\rho g V_{heavy}}{2L^\star} - \rho^\star g \frac{V_{light} - 2L^{\star\star} D}{2(L^{\star\star} - L^\star)}$$
$$- \left\{ \frac{\rho g}{2L^\star} + \frac{\rho^\star g}{2(L^{\star\star} - L^\star)} \right\} V - (\rho - \rho^\star) g Z$$
$$= \gamma \frac{d}{dx} \frac{Z_x}{(1 + Z_x^2)^{\frac{1}{2}}} \tag{2.11}$$

Thus, assuming the drop is pinned at $x = \pm L$, we have $Z = 0$ there and we can solve for the drop shape, $Z(x)$, as a function of the input pressure, p, and then obtain the drop volume, V, as

$$V = -\int_{-L}^{L} Z(x) dx \tag{2.12}$$

We scale x and Z by L and write our drop shape problem in scaled variables. It is

$$P - B_* V - BZ = \frac{d}{dx} \frac{Z_x}{(1 + Z_x^2)^{\frac{1}{2}}} \tag{2.13}$$

where $Z = 0$ at $x = \pm 1$, where we have

$$P = \frac{L}{\gamma} \left\{ p - p_{ref} + \rho g \frac{V_{heavy}}{2L^\star} - \rho^\star g \frac{V_{light} - 2L^{\star\star} D}{2(L^{\star\star} - L^\star)} \right\} \tag{2.14}$$

and

$$V = -\int_{-1}^{1} Z dx \tag{2.15}$$

and where B and B_*, viz.,

$$B = \frac{(\rho - \rho^\star) g L^2}{\gamma} \tag{2.16}$$

and

$$B_* = \frac{L}{\gamma} \{ \frac{\rho g}{2L^\star} + \frac{\rho^\star g}{2(L^{\star\star} - L^\star)} \} L^2 \tag{2.17}$$

are non-negative inputs depending on the design of an experiment and on the fluids being used. By setting $L^\star$ large and $L^{\star\star} - L^\star$ large we can make B_* as small as we like. If $\rho^\star$ is near zero we have $B > B_*$. If $\rho^\star$ is near ρ we have $B < B_*$.

2.3 Solving the Drop Shape Equation

To solve the drop shape equation, viz.,

$$P - B_* V - BZ = \frac{d}{dx} \frac{Z_x}{(1 + Z_x^2)^{\frac{1}{2}}} \tag{2.18}$$

for the drop shape, $Z(x)$, where $Z = 0$ at $x = \pm 1$ and $V = -\int_{-1}^{1} Z dx$, we first set the values of B and B_*. Then for each P we obtain $Z(x)$ and V and we plot our solution as a V vs P curve where at each point along the curve we have $Z(x)$. To do this we hold B fixed and set B_* to zero. Then the drop shape is the solution to the problem

$$P - BZ = \frac{d}{dx} \frac{Z_x}{(1 + Z_x^2)^{\frac{1}{2}}} \tag{2.19}$$

where $Z = 0$ at $x = \pm 1$ and $V = -\int_{-1}^{1} Z dx$. This is our pressure control problem at $B_* = 0$. It is also our volume control problem and we already have its solutions. Thus, say, at $B = \frac{3}{4}\pi^2$ we have the V vs P curve drawn in Fig. 2.3 where we have the drop shape, $Z(x)$, at each point along the curve. These drop shapes solve the pressure control problem not only at $B_* = 0$, Eq. (2.19), but also at B_* greater than zero, Eq. (2.18).

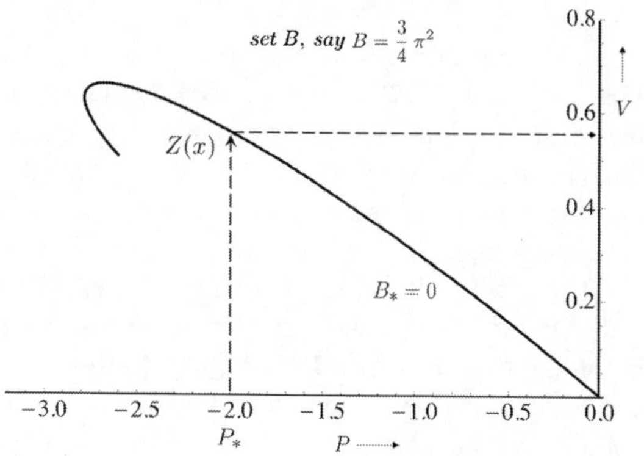

Fig. 2.3 The V vs P curve for a drop under volume control. It is also the V *vs* P curve for a drop under pressure control at $B_* = 0$

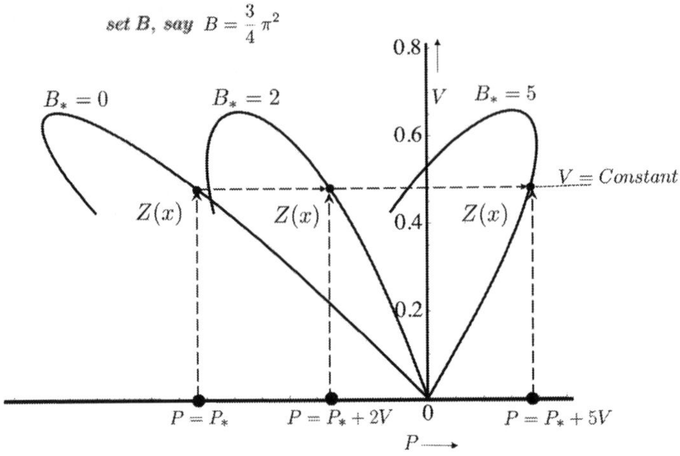

Fig. 2.4 Under pressure control, upon increasing B_*, drop shapes and their volumes do not change so long as P changes according to $P = P_* + B_* V$

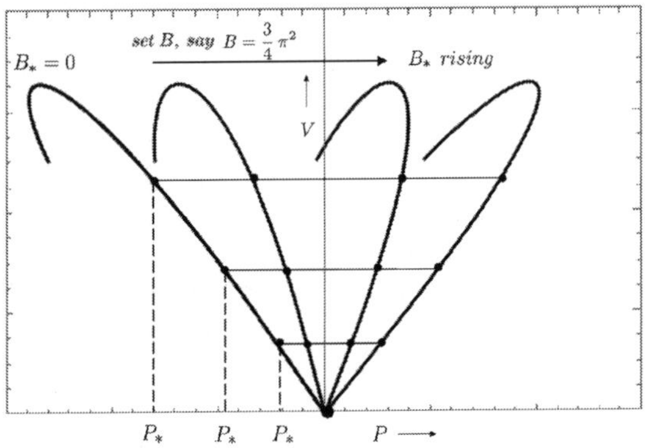

Fig. 2.5 V vs P curves at $B_* > 0$. They are derived from the V vs P curve at $B_* = 0$

To see that this is so, we set $P - B_* V$ in Eq. (2.18) to P_*, whereupon we have Eq. (2.19) at $P = P_*$. Then we view Eq. (2.19) as a volume control problem, set the volume, V, and obtain the drop shape, $Z(x)$, and the pressure P_*. The result is this: the drop shape, $Z(x)$, that we have obtained upon solving the volume control problem, solves the pressure control problem, first at $B_* = 0$ and at the pressure P_*, then at any input $B_* > 0$ and at the pressure P, where $P = P_* + B_* V$, cf., Fig. 2.4.

Thus holding B fixed and having $Z(x)$ and V at $B_* = 0$ and at $P = P_*$, $Z(x)$ and V do not change as B_* rises from zero so long as P is adjusted along the way according to $P = P_* + B_* V$. We can do this at many V's and obtain Fig. 2.5. All drop shapes under pressure control obtain by solving the drop shape problem under volume control.

2.4 Does the V vs P Curve Start Out in the First or the Second Quadrant?

Our V vs P curves start out at $P = 0$, $Z(x) = 0$ and $V = 0$. In Fig. 2.5 we see that if the V vs P curve under volume control, i.e., the V vs P curve under pressure control at $B_* = 0$, lies in the second quadrant, then, for small values of B_*, the V vs P curve under pressure control also lies in the second quadrant. But at large values of B_* it starts out in the first quadrant where the pressure rises from zero, passes through its greatest value, denoted P_{nose}, at the nose of the curve before turning toward the V axis.

We wish to find out, given B and B_*, whether our V vs P curve starts out in the first or the second quadrant, i.e., must P increase or decrease from zero in order that V increase? To do this we derive a formula for $\dot{V}$, where $\dot{V} = \frac{dV}{dP}$. Upon differentiating Eq. (2.18) with respect to P, we obtain

$$1 + B_* \int_{-1}^{1} \dot{Z} dx - B\dot{Z} = \frac{d}{dx} \frac{\dot{Z}_x}{(1 + Z_x^2)^{\frac{3}{2}}} \tag{2.20}$$

where $\dot{Z} = 0$ at $x = \pm 1$. Then at $P = 0$, $Z(x) = 0$, $V = 0$ we have

$$1 + B_* \int_{-1}^{1} \dot{Z} dx - B\dot{Z} = \frac{d^2}{dx^2} \dot{Z} \tag{2.21}$$

and assuming $Z(x)$ is even in x we have $\dot{Z}(x)$ even in x whereupon we have

$$\dot{Z} = A \cos \sqrt{B} x + \frac{1 + B_* \int_{-1}^{1} \dot{Z} dx}{B} \tag{2.22}$$

Obtaining A so that $\dot{Z}$ is zero at $x = \pm 1$ we have

$$\dot{V} = -\int_{-1}^{1} \dot{Z} dx = \frac{\frac{2}{B}\left(\frac{\tan\sqrt{B}}{\sqrt{B}} - 1\right)}{1 + B_* \frac{2}{B}\left(\frac{\tan\sqrt{B}}{\sqrt{B}} - 1\right)} \tag{2.23}$$

For $0 < B < \frac{1}{4}\pi^2$ we sketch $\frac{2}{B}\left(\frac{\tan\sqrt{B}}{\sqrt{B}} - 1\right)$ vs B in Fig. 2.6 whereupon, for all B_*, $\dot{V}$ is positive and our V vs P curve starts out in the first quadrant.

For $\frac{1}{4}\pi^2 < B < \pi^2$ we define $N(B) > 0$ by

$$N(B) = -\frac{2}{B}\left(\frac{\tan\sqrt{B}}{\sqrt{B}} - 1\right) \tag{2.24}$$

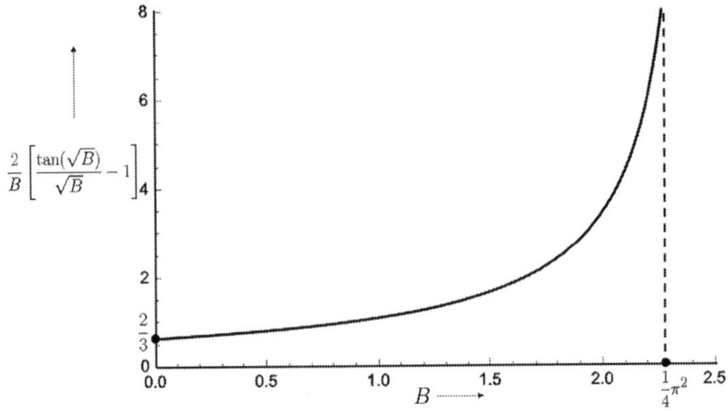

Fig. 2.6 $\frac{2}{B}\left(\frac{\tan\sqrt{B}}{\sqrt{B}}-1\right)$ *vs* $B, 0 < B < \frac{1}{4}\pi^2$

and we write

$$\overset{\bullet}{V} = \frac{-N(B)}{1 - B_* N(B)} \tag{2.25}$$

Then for $B_* = 0$, $\overset{\bullet}{V}$ is negative and our V vs P curve starts out in the second quadrant. It starts out in the second quadrant for $0 < B_* < \frac{1}{N(B)}$. At $B_* = \frac{1}{N(B)}$ it starts upward along the V axis. Then for all $B_* > \frac{1}{N(B)}$ it starts out in the first quadrant. By solving

$$B_* = \frac{1}{N(B)} \tag{2.26}$$

for B and plotting B vs B_* we have Fig. 2.7.

Our result is this: our V vs P curves start out in the first quadrant for $0 < B < \frac{1}{4}\pi^2$ no matter the value of B_*. For $\frac{1}{4}\pi^2 < B < \pi^2$ they start out in the second quadrant at $B_* = 0$. Then as B_* increases they start out in the second quadrant until B_* reaches $\frac{1}{N(B)}$. Thereafter they start out in the first quadrant.

2.5 Drawing the V *vs* P Curve at Input Values of B and B_*

We set the value of B. Then, having solved the volume control problem, at $B_* = 0$ we have the V vs P curve and the drop shapes along this curve. If B lies on $\left(0, \frac{1}{4}\pi^2\right)$ this curve starts out in the first quadrant before crossing the V axis. If B lies on $\left(\frac{1}{4}\pi^2, \pi^2\right)$ it lies entirely in the second quadrant.

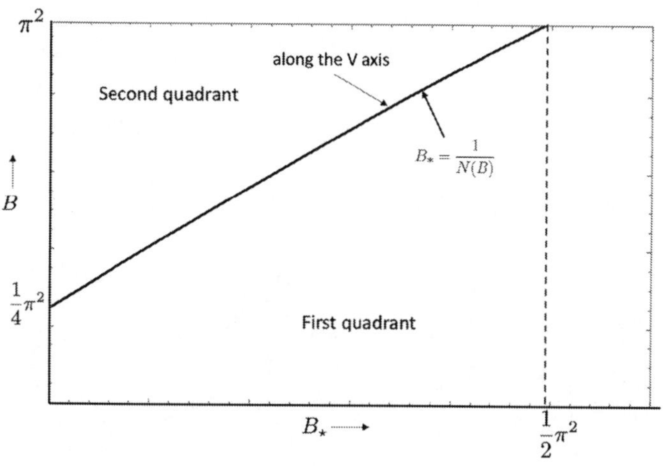

Fig. 2.7 V vs P curves start out in the first quadrant, along the V- axis or in the second quadrant according to the values of B and B_*

In Sect. 2.3 we set a drop shape $Z(x)$ at pressure P_* at $B_* = 0$. Then we held $Z(x)$ fixed, and therefore V, and advanced B_*. At each B_* we obtained P via $P = P_*(V) + B_* V$ such that the point (P, V) lies on the V vs P curve at B_*. At all these B_*'s the drop shape remains $Z(x)$.

Now we set B_* and obtain the V vs P curve at input values of B and B_*. Having set P, the drop shape, and so too its volume, is our output. The drop shape does not remain fixed as we increase P and move along the V vs P curve.

First we set B on the interval $\left(\frac{1}{4}\pi^2, \pi^2\right)$, whereupon the V vs P curve at $B_* = 0$ lies in the second quadrant. Then we set B_*. We begin by setting B_* large enough that the V vs P curve at B_*, the curve we are looking for, starts out in the first quadrant. To obtain our curve we set the pressure, denoted P, and draw the straight line V vs P_*, viz., $V = -\frac{1}{B_*} P_* + \frac{1}{B_*} P$. Then where it intersects the V vs P curve at $B_* = 0$ we have two drop shapes, and their volumes, along the V vs P curve at B_*, cf., Fig. 2.8.

The V vs P curve obtains by drawing all possible straight lines of slope $-\frac{1}{B_*}$, cf., Fig. 2.9. The greatest pressure, denoted P_{nose}, obtains where the line of slope $-\frac{1}{B_*}$ is just tangent to the V vs P curve at $B_* = 0$. The volume at the point where the line is tangent to the curve is denoted V_{nose}. The point (P_{nose}, V_{nose}) is a turning point. Thus given B_* we have the turning point (P_{nose}, V_{nose}), cf., Fig. 2.9.

If we do not set B_* large enough, the V vs P curve at B_* will lie in the second quadrant and the greatest pressure will be $P = 0$, ordinarily not a nose., cf., Fig. 2.10.

For B_*'s large enough that the V vs P curve at B_* starts out in the first quadrant we can obtain the greatest pressure, P_{nose}, and its volume, V_{nose}, along each V vs P curve by drawing a family of tangents, one for each B_*, cf., Fig. 2.11.

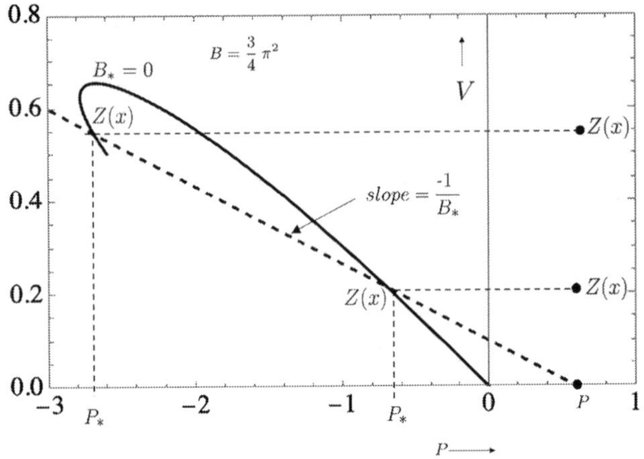

Fig. 2.8 Obtaining two points on the V vs P curve at B_*

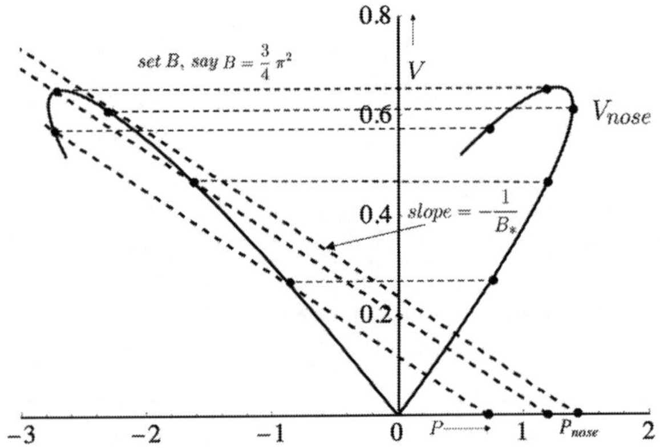

Fig. 2.9 V vs P curve at B_* Obtaining P_{nose}, V_{nose}

The least value of B_* at which we have a turning point is the value of B_* such that $-\frac{1}{B_*}$ is equal to $\dot{V}(B_* = 0, V = 0) = -N(B)$, i.e., $B_* = \frac{1}{N(B)}$, cf., Fig. 2.12.

Having set the value of $B \epsilon \left(\frac{1}{4}\pi^2, \pi^2\right)$, at each value of B_* greater than $\frac{1}{N(B)}$ we have a greatest pressure, denoted P_{nose}, which occurs at the volume V denoted V_{nose} along the V vs P curve at B_*. Thus, on the $V - B_*$ plane, we can plot the curve along which the pressure takes on its greatest value, P_{nose}, where $P_{nose} = P_*(V_{nose}) + B_* V_{nose}$, cf., Fig. 2.13. The points along this curve, which we call turning points, will turn out to be critical points upon advancing P upward from zero at given values of B and B_*.

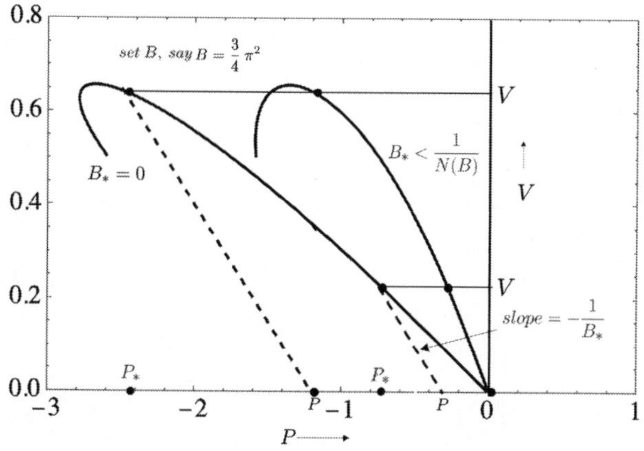

Fig. 2.10 The V vs P curve at a small value of B_*, $B_* < \frac{1}{N(B)}$

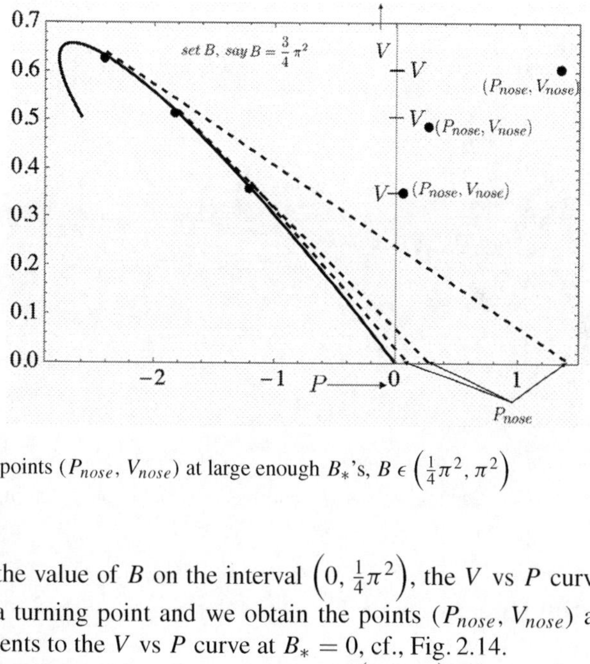

Fig. 2.11 The points (P_{nose}, V_{nose}) at large enough B_*'s, $B \in \left(\frac{1}{4}\pi^2, \pi^2\right)$

If we set the value of B on the interval $\left(0, \frac{1}{4}\pi^2\right)$, the V vs P curve at $B_* = 0$ already has a turning point and we obtain the points (P_{nose}, V_{nose}) as before, by drawing tangents to the V vs P curve at $B_* = 0$, cf., Fig. 2.14.

Thus we have the turning point curve at $B \in \left(0, \frac{1}{4}\pi^2\right)$ drawn in Fig. 2.15, where the arrow points to the volume at the turning point at $B_* = 0$. At $B = \frac{1}{4}\pi^2$ the curve starts at $B_* = 0$, $V = 0$.

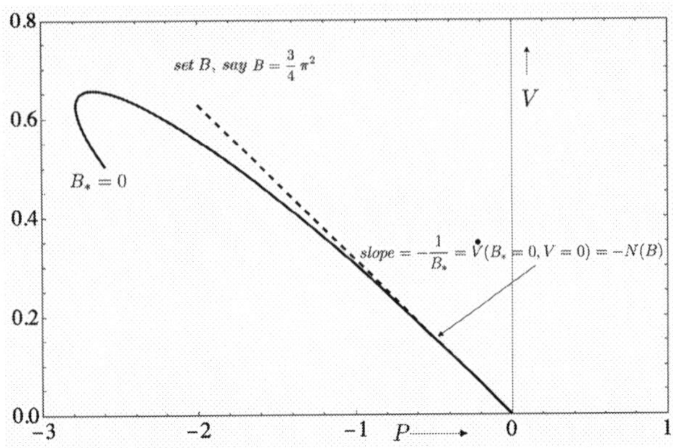

Fig. 2.12 The least B_* at which we have a turning point, $B \in \left(\frac{1}{4}\pi^2, \pi^2\right)$

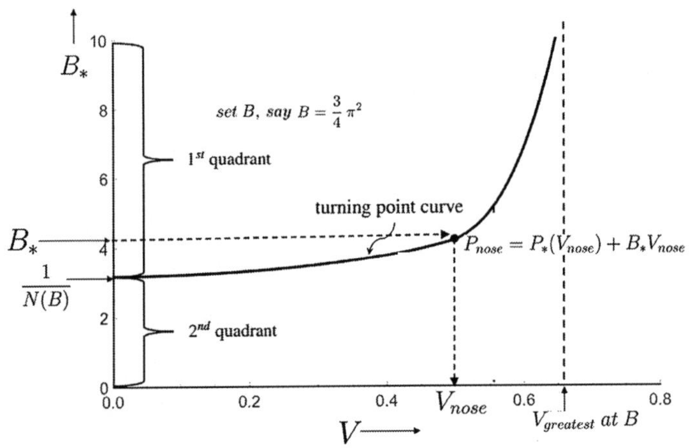

Fig. 2.13 Given $B \in \left(\frac{1}{4}\pi^2, \pi^2\right)$, at each $B_* > \frac{1}{N(B)}$ we have the volume , V_{nose}, at which the pressure reaches its greatest value, P_{nose}, at B_*. This is a curve of turning points in pressure on the $V - B_*$ plane

2.6 Potential Energy

At input values of B and B_* we wish to know if a drop of shape $Z(x)$ at the point (P, V) is stable to small displacements. To find out we turn to Rayleigh's work principle which tells us that a drop is stable if and only if its potential energy increases for all small displacements.

Thus our first job is to obtain the potential energy of a drop of shape $Z(x)$. Turning our attention to Fig. 2.2, we begin by working in unscaled variables. We set the pressure, hold it constant, then give our drop a small displacement.

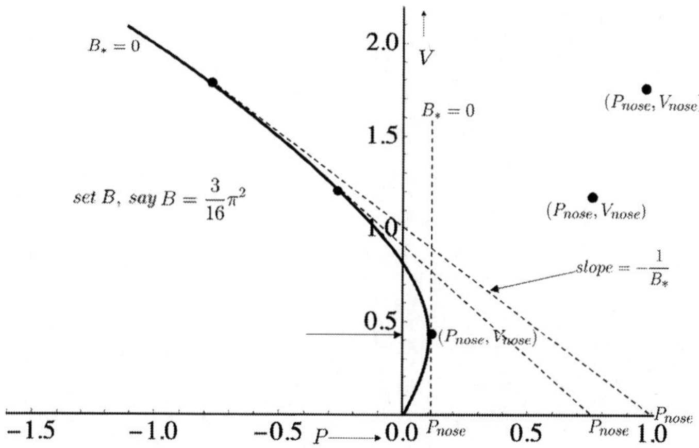

Fig. 2.14 Given $B \epsilon \left(0, \frac{1}{4}\pi^2\right)$, points (P_{nose}, V_{nose}) for B_*'s ≥ 0

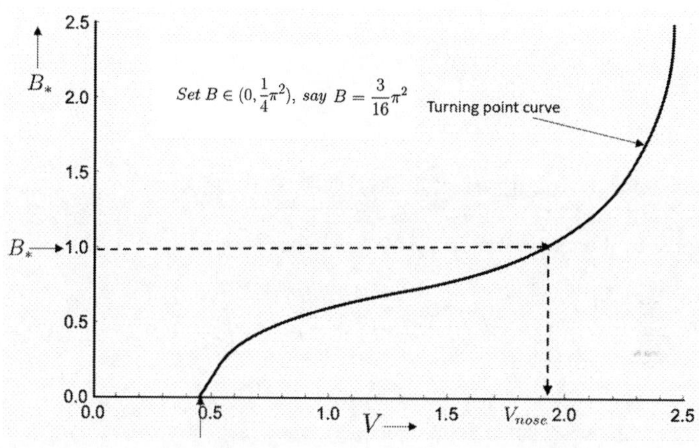

Fig. 2.15 Given $B \epsilon \left(0, \frac{1}{4}\pi^2\right)$, at each B_* we have the volume, V_{nose}, at which the pressure reaches its greatest value, P_{nose} at B_*. This is a curve of turning points in pressure on the $V - B_*$ plane

The contributions to the potential energy are:
surface:

$$\int_{-L}^{L} \gamma(1 + Z_x{}^2)^{\frac{1}{2}} dx \tag{2.27}$$

gravity: heavy fluid

$$\int_{-L^\star}^{-L} dx \int_0^h dz\, \rho g z + \int_{-L}^{L} dx \int_{Z(x)}^{h} dz\, \rho g z + \int_{L}^{L^\star} dx \int_0^h dz\, \rho g z \qquad (2.28)$$
$$= \rho g L^\star h^2 - \rho g \frac{1}{2} \int_{-L}^{L} dx\, Z^2$$

gravity: light fluid

$$\int_{-L^{\star\star}}^{-L^\star} dx \int_{-D}^{h^\star} dz\, \rho^\star g z + \int_{-L^\star}^{-L} dx \int_{-D}^{0} dz\, \rho^\star g z + \int_{-L}^{L} dx \int_{-D}^{Z(x)} dz\, \rho^\star g z$$
$$+ \int_{L}^{L^\star} dx \int_{-D}^{0} dz\, \rho^\star g z + \int_{L^\star}^{L^{\star\star}} dx \int_{-D}^{h^\star} dz\, \rho^\star g z$$
$$= \rho^\star g (L^{\star\star} - L^\star) h^{\star 2} + \rho^\star g \frac{1}{2} \int_{-L}^{L} dx\, Z^2(x) \qquad (2.29)$$

where we have dropped the term $\rho^\star g\, L^{\star\star}(-D)^2$
pressure:

$$p\, 2L^\star\, h + p_{ref}\, 2(L^{\star\star} - L^\star) h^\star \qquad (2.30)$$

Thus the potential energy of a drop having shape $Z(x)$ is

$$PE(Z) = \int_{-L}^{L} \gamma\, (1 + Z_x{}^2)^{\frac{1}{2}}\, dx - (\rho - \rho^\star) g \int_{-L}^{L} \frac{1}{2} Z^2\, dx + \rho g\, L^\star\, h^2$$
$$(2.31)$$
$$+ \rho^\star g (L^{\star\star} - L^\star) h^{\star 2} + p\, 2L^\star\, h + p_{ref}\, 2(L^{\star\star} - L^\star) h^\star$$

where

$$2L^\star\, h = V_{heavy} - V$$
$$2(L^{\star\star} - L^\star) h^\star = V_{light} - 2LD + V \qquad (2.32)$$

and

$$V = -\int_{-L}^{L} Z(x)\, dx$$

Now we write

$$Z(x,\epsilon) = Z_0(x) + \epsilon\, Z_1(x) + \frac{1}{2}\,\epsilon^2\, Z_2(x) \tag{2.33}$$

where $Z_0(x)$ is the shape of an equilibrium drop and $\epsilon\, Z_1(x) + \frac{1}{2}\,\epsilon^2\, Z_2(x)$ denotes a small displacement and where $Z|_{\epsilon=0} = Z_0(x)$, $\frac{dZ}{d\epsilon}\big|_{\epsilon=0} = Z_1(x)$, etc.

Our aim is to derive the drop shape equation by setting $\left.\frac{dPE}{d\epsilon}\right|_{\epsilon=0}$ to zero and then obtain a formula for $\left.\frac{d^2PE}{d\epsilon^2}\right|_{\epsilon=0}$.

Upon differentiating Eq. (2.31) with respect to ϵ we have

$$\begin{aligned}\frac{dPE}{d\epsilon} &= \int_{-L}^{L} \gamma\, \frac{1}{(1+{Z_x}^2)^{\frac{1}{2}}}\, Z_x\, \frac{dZ_x}{d\epsilon}\, dx - (\rho-\rho^\star) g \int_{-L}^{L} Z\, \frac{dZ}{d\epsilon}\, dx \\ &+ \rho g\, L^\star\, 2h \frac{dh}{d\epsilon} + \rho^\star g (L^{\star\star} - L^\star)\, 2h^\star\, \frac{dh^\star}{d\epsilon} \\ &+ p\, 2L^\star\, \frac{dh}{d\epsilon} + p_{ref}\, 2(L^{\star\star} - L^\star)\, \frac{dh^\star}{d\epsilon}\end{aligned} \tag{2.34}$$

where

$$2L^\star\, \frac{dh}{d\epsilon} = -\frac{dV}{d\epsilon} = \int_{-L}^{L} \frac{dZ}{d\epsilon}\, dx \tag{2.35}$$

and

$$2(L^{\star\star} - L^\star)\, \frac{dh^\star}{d\epsilon} = \frac{dV}{d\epsilon} = -\int_{-L}^{L} \frac{dZ}{d\epsilon}\, dx \tag{2.36}$$

whereupon we have

$$\begin{aligned}\frac{dPE}{d\epsilon} = \int_{-L}^{L} \Bigg\{ &\gamma \frac{1}{(1+{Z_x}^2)^{\frac{1}{2}}}\, Z_x\, \frac{dZ_x}{d\epsilon} - (\rho-\rho^\star) g\, Z \frac{dZ}{d\epsilon} \\ &+ \rho g h\, \frac{dZ}{d\epsilon} - \rho^\star g h^\star\, \frac{dZ}{d\epsilon} + p \frac{dZ}{d\epsilon} - p_{ref} \frac{dZ}{d\epsilon} \Bigg\}\, dx\end{aligned} \tag{2.37}$$

and

$$\begin{aligned}\left.\frac{dPE}{d\epsilon}\right|_{\epsilon=0} = \int_{-L}^{L} \Bigg\{ &\gamma \frac{Z_{0x}}{(1+{Z_{0x}}^2)^{\frac{1}{2}}}\, Z_{1x} - (\rho-\rho^\star) g\, Z_0 Z_1 \\ &+ \rho g h_0\, Z_1 - \rho^\star g {h_0}^\star\, Z_1 + p Z_1 - p_{ref} Z_1 \Bigg\}\, dx\end{aligned} \tag{2.38}$$

Then, upon integrating by parts and accounting for $Z_1 = 0$ at $\pm L$, we have

$$\left.\frac{dPE}{d\epsilon}\right|_{\epsilon=0} = \int_{-L}^{L} \left\{ -\gamma \frac{d}{dx} \frac{Z_{0x}}{(1 + {Z_{0x}}^2)^{\frac{1}{2}}} - (\rho - \rho^\star) g\, Z_0 \right. \\ \left. +\rho g h_0 - \rho^\star g {h_0}^\star + p - p_{ref} \right\} Z_1\, dx \tag{2.39}$$

where in order that $\left.\frac{dPE}{d\epsilon}\right|_{\epsilon=0} = 0$ no matter $Z_1(x)$ we obtain

$$p - p_{ref} + \rho g h_0 - \rho^\star g {h_0}^\star - (\rho - \rho^\star) g\, Z_0 = \gamma \frac{d}{dx} \frac{Z_{0x}}{(1 + {Z_{0x}}^2)^{\frac{1}{2}}} \tag{2.40}$$

This is Eq. (2.8) for the equilibrium drop shape $Z_0(x)$ given the value of the input pressure p, where h_0 and $h_0^\star$ depend on $Z_0(x)$ via $\int_{-L}^{L} Z_0(x)dx$.

We turn now to $\left.\frac{d^2PE}{d\epsilon^2}\right|_{\epsilon=0}$ and account at first for terms in which only $Z_2(x)$ appears. Thus we have

$$\frac{d^2PE}{d\epsilon^2} = \int_{-L}^{L} \left\{ \gamma \frac{1}{(1 + {Z_x}^2)^{\frac{1}{2}}} Z_x \frac{d^2 Z_x}{d\epsilon^2} - (\rho - \rho^\star) g\, Z \frac{d^2 Z}{d\epsilon^2} \right. \\ \left. +\rho g h \frac{d^2 Z}{d\epsilon^2} - \rho^\star g h^\star \frac{d^2 Z}{d\epsilon^2} + p \frac{d^2 Z}{d\epsilon^2} - p_{ref} \frac{d^2 Z}{d\epsilon^2} \right\} dx \tag{2.41}$$

and

$$\left.\frac{d^2PE}{d\epsilon^2}\right|_{\epsilon=0} = \int_{-L}^{L} \left\{ \gamma \frac{Z_{0x}}{(1 + {Z_{0x}}^2)^{\frac{1}{2}}} Z_{2x} - (\rho - \rho^\star) g\, Z_0 Z_2 \right. \\ \left. + \rho g h_0\, Z_2 - \rho^\star g {h_0}^\star\, Z_2 + p Z_2 - p_{ref} Z_2 \right\} dx \equiv RHS \tag{2.42}$$

Then, integrating by parts and accounting for $Z_2 = 0$ at $\pm L$, we have

$$RHS = \int_{-L}^{L} \left\{ -\gamma \frac{d}{dx} \frac{Z_{0x}}{(1 + {Z_{0x}}^2)^{\frac{1}{2}}} - (\rho - \rho^\star) g\, Z_0 \right. \\ \left. +\rho g h_0 - \rho^\star g {h_0}^\star + p - p_{ref} \right\} Z_2\, dx \tag{2.43}$$

and this is zero no matter $Z_2(x)$ due to $Z_0(x)$ being an equilibrium drop shape satisfying

$$p - p_{ref} + \rho g h_0 - \rho^\star g h_0{}^\star - (\rho - \rho^\star) g\, Z_0 = \gamma\, \frac{d}{dx} \frac{Z_{0x}}{(1 + Z_{0x}{}^2)^{\frac{1}{2}}}$$

Thus $\left.\frac{d^2 PE}{d\epsilon^2}\right|_{\epsilon=0}$ does not depend on $Z_2(x)$ and we turn to the remaining part of $\frac{d^2 PE}{d\epsilon^2}$. It is

$$\frac{d^2 PE}{d\epsilon^2} = \int_{-L}^{L} \left[\gamma\, \frac{d}{d\epsilon} \left\{ \frac{1}{(1 + Z_x{}^2)^{\frac{1}{2}}}\, Z_x \right\} \frac{dZ_x}{d\epsilon} + \left(-\,(\rho - \rho^\star) g\, \frac{dZ}{d\epsilon} \right) \frac{dZ}{d\epsilon} \right. \tag{2.44}$$

$$\left. + \left(\rho g \frac{dh}{d\epsilon} - \rho^\star g \frac{dh^\star}{d\epsilon} \right) \frac{dZ}{d\epsilon} \right] dx$$

where

$$\rho g \frac{dh}{d\epsilon} = \frac{\rho g}{2L^\star} \int_{-L}^{L} \frac{dZ}{d\epsilon}\, dx \tag{2.45}$$

$$-\,\rho^\star g \frac{dh^\star}{d\epsilon} = \frac{\rho^\star g}{2(L^{\star\star} - L^\star)} \int_{-L}^{L} \frac{dZ}{d\epsilon}\, dx \tag{2.46}$$

and

$$\frac{d}{d\epsilon} \left\{ \frac{1}{(1 + Z_x^2)^{\frac{1}{2}}} Z_x \right\} = \frac{1}{(1 + Z_x{}^2)^{\frac{3}{2}}} \frac{dZ_x}{d\epsilon} \tag{2.47}$$

whereupon we have

$$\frac{d^2 PE}{d\epsilon^2} = \int_{-L}^{L} \left[\left\{ \gamma\, \frac{1}{(1 + Z_x{}^2)^{\frac{3}{2}}} \frac{dZ_x}{d\epsilon} \right\} \frac{dZ_x}{d\epsilon} + \left\{ -\,(\rho - \rho^\star) g\, \frac{dZ}{d\epsilon} \right. \right. \tag{2.48}$$

$$\left. \left. + \left(\frac{\rho g}{2L^\star} - \frac{\rho^\star g}{2(L^{\star\star} - L^\star)} \right) \int_{-L}^{L} \frac{dZ}{d\epsilon}\, dx \right\} \frac{dZ}{d\epsilon} \right] dx$$

and, therefore,

$$\left.\frac{d^2 PE}{d\epsilon^2}\right|_{\epsilon=0} = \int_{-L}^{L} \left[\gamma\, \frac{Z_{1x}}{(1 + Z_{0x}{}^2)^{\frac{3}{2}}}\, Z_{1x} + \left\{ -\,(\rho - \rho^\star) g\, Z_1 \right. \right. \tag{2.49}$$

$$\left. \left. + \left(\frac{\rho g}{2L^\star} - \frac{\rho^\star g}{2(L^{\star\star} - L^\star)} \right) \int_{-L}^{L} Z_1\, dx \right\} Z_1 \right] dx$$

Then integrating by parts and setting Z_1 to zero at $x = \pm L$ we have

$$\left.\frac{d^2 PE}{d\epsilon^2}\right|_{\epsilon=0} = \int_{-L}^{L} \left\{ -\gamma \frac{d}{dx} \frac{Z_{1x}}{(1 + Z_{0x}{}^2)^{\frac{3}{2}}} - (\rho - \rho^\star) g\, Z_1 \right. \tag{2.50}$$

$$\left. + \left(\frac{\rho g}{2L^\star} - \frac{\rho^\star g}{2(L^{\star\star} - L^\star)} \right) \int_{-L}^{L} Z_1\, dx \right\} Z_1\, dx$$

and scaling Z and x by L, we have, in scaled variables,

$$\left.\frac{d^2 PE}{d\epsilon^2}\right|_{\epsilon=0} = -\int_{-1}^{1} \left\{ \frac{d}{dx} \frac{Z_{1x}}{(1 + Z_{0x}{}^2)^{\frac{3}{2}}} + B Z_1 - B_* \int_{-1}^{1} Z_1 dx \right\} Z_1\, dx \tag{2.51}$$

where

$$B = \frac{(\rho - \rho^*) g\, L^2}{\gamma}$$

and

$$B_* = \frac{L}{\gamma} \left(\frac{\rho g}{2L^\star} + \frac{\rho^\star g}{2(L^{\star\star} - L^\star)} \right) L^2$$

Now $Z_0(x)$ is the shape of an equilibrium drop, viz., at $Z_0(x)$ we have $\left.\frac{dPE}{d\epsilon}\right|_{\epsilon=0} = 0$. It is a stable drop if and only if $\left.\frac{d^2 PE}{d\epsilon^2}\right|_{\epsilon=0}$ is positive for all displacements, $Z_1(x)$. To obtain our displacements we solve an eigenvalue problem. It is

$$B_* \int_{-1}^{1} \psi\, dx - B\psi = \frac{d}{dx} \frac{\psi_x}{(1 + Z_{0x}^2)^{\frac{3}{2}}} + \lambda^2 \psi \tag{2.52}$$

where $\psi = 0$ at $x = \pm 1$. Then setting Z_1 to any eigenfunction ψ we have

$$\boxed{\left.\frac{d^2 PE}{d\epsilon^2}\right|_{\epsilon=0} = \lambda^2 \int_{-1}^{1} \psi^2\, dx} \tag{2.53}$$

and we see that a drop of shape $Z_0(x)$ is stable if and only if all eigenvalues at $Z_0(x)$ are positive, i.e., given a drop having the shape $Z_0(x)$, where $\left.\frac{dPE}{d\epsilon}\right|_{\epsilon=0} = 0$, for all displacements $Z_1(x)$, we have

$$\Delta PE = \epsilon \left.\frac{dPE}{d\epsilon}\right|_{\epsilon=0} + \frac{1}{2} \epsilon^2 \left.\frac{d^2 PE}{d\epsilon^2}\right|_{\epsilon=0} = \frac{1}{2} \epsilon^2 \left.\frac{d^2 PE}{d\epsilon^2}\right|_{\epsilon=0} \tag{2.54}$$

2.7 Critical Points

Having set the inputs, B and B_*, we have a V vs P curve and at each point along the curve we have the drop shape, $Z(x)$, at the pressure, P. We wish to know where, as we advance along the curve, the drop has become critical, i.e., where, along the curve, the drop is just about to break. To find out whether or not a drop has reached a critical pressure, we solve an eigenvalue problem.

Our eigenvalue problem is

$$B_* \int_{-1}^{1} \psi dx - B\psi = \frac{d}{dx} \frac{\psi_x}{(1 + Z_x^2)^{\frac{3}{2}}} + \lambda^2 \psi \tag{2.55}$$

where $\psi = 0$ at $x = \pm 1$. The eigenvalues are denoted λ^2, the eigenfunctions are denoted ψ. Then, due to $Z(x)$ being an even function of x, the eigenfunctions are either even or odd functions of x and we call eigenvalues even or odd according as the eigenfunctions are even or odd.

Thus to locate critical drop shapes we start at $P = 0$ where we assume that all the eigenvalues are positive. Then as we advance P and move along the V vs P curve the eigenvalues move to the left and we solve for the least eigenvalue at each point along the way. Doing this we obtain the value of P where the least eigenvalue first reaches zero. This is the critical value of P and the drop is just about to break. We obtain the solution to our eigenvalue problem in end note 2.12.1.

Before going on, we derive the result that all eigenvalues are positive at $P = 0$, where $Z(x) = 0$, if and only if the V vs P curve starts out in the first quadrant. The eigenvalue problem at $Z(x) = 0$ is

$$B_* \int_{-1}^{1} \psi dx - B\psi = \frac{d^2\psi}{dx^2} + \lambda^2 \psi \tag{2.56}$$

where $\psi = 0$ at $x = \pm 1$. Its odd solutions are

$$\psi = A \sin(\lambda^2 + B)^{\frac{1}{2}} x \tag{2.57}$$

where

$$0 = A \sin(\lambda^2 + B)^{\frac{1}{2}} \tag{2.58}$$

and where, to have A not zero, we must have

$$\lambda^2 = -B + n^2\pi^2, \quad n = 1, 2, \ldots \tag{2.59}$$

Thus all the odd eigenvalues (and so too all even eigenvalues except possibly the least even eigenvalue) are positive for all B's less than π^2, no matter the value of B_*.

Turning to the even solutions, where now λ^2 denotes the least even eigenvalue, we have

$$\psi = A\cos(\lambda^2 + B)^{\frac{1}{2}} x + \frac{B_*}{\lambda^2 + B} \int_{-1}^{1} \psi dx \tag{2.60}$$

where

$$\int_{-1}^{1} \psi dx = A \frac{2\sin(\lambda^2 + B)^{\frac{1}{2}}}{(\lambda^2 + B)^{\frac{1}{2}}} + \frac{2B_*}{\lambda^2 + B} \int_{-1}^{1} \psi dx \tag{2.61}$$

and, therefore, where

$$\frac{B_*}{\lambda^2 + B} \int_{-1}^{1} \psi dx = A \frac{\frac{2B_*}{\lambda^2+B}}{1 - \frac{2B_*}{\lambda^2+B}} \frac{\sin(\lambda^2 + B)^{\frac{1}{2}}}{(\lambda^2 + B)^{\frac{1}{2}}} \tag{2.62}$$

Thus we have

$$\psi = A \left\{ \cos(\lambda^2 + B)^{\frac{1}{2}} x + \frac{\frac{2B_*}{\lambda^2+B}}{1 - \frac{2B_*}{\lambda^2+B}} \frac{\sin(\lambda^2 + B)^{\frac{1}{2}}}{(\lambda^2 + B)^{\frac{1}{2}}} \right\} \tag{2.63}$$

and, due to $\psi = 0$ at $x = \pm 1$, we have A not zero, and therefore ψ not zero, if and only if we have

$$\cos(\lambda^2 + B)^{\frac{1}{2}} + \frac{\frac{2B_*}{\lambda^2+B}}{1 - \frac{2B_*}{\lambda^2+B}} \frac{\sin(\lambda^2 + B)^{\frac{1}{2}}}{(\lambda^2 + B)^{\frac{1}{2}}} = 0 \tag{2.64}$$

which we rewrite as

$$B_* = \frac{1}{N(\lambda^2 + B)}, \quad B_* \geq 0 \tag{2.65}$$

where

$$N(x) = -\frac{2}{x} \left(\frac{\tan\sqrt{x}}{\sqrt{x}} - 1 \right) \tag{2.66}$$

This is an equation for λ^2 as a function of B and B_*.

If B lies on $\left(0, \frac{1}{4}\pi^2\right)$, $N(B)$ is negative whereupon λ^2 must be positive, i.e., $\lambda^2 + B$ must be greater than $\frac{1}{4}\pi^2$, no matter the value of B_*.

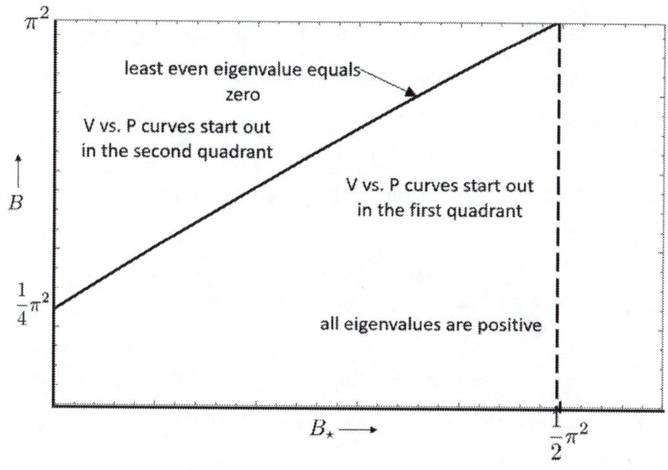

Fig. 2.16 *V* vs *P* curves start out stable if and only if they start out in the first quadrant

If B lies on $\left(\frac{1}{4}\pi^2, \pi^2\right)$ we have λ^2 equal to zero along the curve obtained by solving

$$B_* = \frac{1}{N(B)} \tag{2.67}$$

for B and plotting B vs B_*. Thus we have Fig. 2.16 where we see that V vs P curves start out stable if and only if they start out in the first quadrant.

2.8 Branch Points, Turning Points

As we advance along the V vs P curve at input values of B and B_*, we expect to arrive at a critical point, a point where the least eigenvalue has just reached zero. We wish to know if we have reached the greatest value of P or if we can continue to increase P and pass through this point. We have reached the greatest possible pressure if we cannot solve for $\dot{Z}(x)$ at P, where $\dot{Z} = \frac{dZ}{dP}$. The $\dot{Z}$ equation obtains on differentiating the Z equation with respect to P. It is

$$1 + B_* \int_{-1}^{1} \dot{Z} dx - B\dot{Z} = \frac{d}{dx} \frac{\dot{Z}_x}{(1 + Z_x^2)^{\frac{3}{2}}} \tag{2.68}$$

where $\overset{\bullet}{Z} = 0$ at $x = \pm 1$. This is a linear non-homogeneous equation and at a critical point the homogeneous $\overset{\bullet}{Z}$ equation has a non-zero solution, viz., the eigenfunction corresponding to the eigenvalue that has just reached zero. Therefore to solve for $\overset{\bullet}{Z}$ at a critical point a solvability condition must be satisfied.

Now the eigenfunction solving the homogeneous $\overset{\bullet}{Z}$ equation is the solution to

$$B_* \int_{-1}^{1} \psi dx - B\psi = \frac{d}{dx} \frac{\psi_x}{(1 + Z_x^2)^{\frac{3}{2}}} \tag{2.69}$$

where $\psi = 0$ at $x = \pm 1$. Thus multiplying the $\overset{\bullet}{Z}$ equation by ψ and the ψ equation by $\overset{\bullet}{Z}$, subtracting and integrating over $(-1, 1)$ we obtain the solvability condition. It is

$$\int_{-1}^{1} \psi dx = 0 \tag{2.70}$$

If ψ is odd, i.e., if the least odd eigenvalue is the first to reach zero, solvability is satisfied, the critical point is a branch point and we can increase P and pass through the critical pressure. But if ψ is even, i.e., the least even eigenvalue is the first to reach zero, it cannot integrate to zero, cf., end note 2.12.1 where we see that all ψ's integrating to zero must be odd ϕ's. Thus solvability is not satisfied, we cannot obtain $\overset{\bullet}{Z}$ and we have reached the greatest value of P. The critical point is then a turning point and we have reached P_{nose}.

We have established an important connection: P_{nose} corresponds to the least even eigenvalue reaching zero. Thus our turning point curves on the $V - B_*$ plane are least-even-eigenvalue-equal-to-zero curves and setting a value of B, say, $\frac{1}{4}\pi^2 < B < \pi^2$, we have Fig. 2.17. What we have learned that we did not know earlier is this: a nose is a critical point where the drop is just about to break and the pattern at breakup is derived from the even eigenfunction corresponding to the least even eigenvalue reaching zero.

Given B, the curve in Fig. 2.17, along which the least even eigenvalue is zero, is the turning point curve, drawn knowing only the V vs P curve at $B_* = 0$, cf., Fig. 2.13. No eigenvalue problem has been solved to find out where the least even eigenvalue first reaches zero.

Turning our attention to the odd eigenvalues, we set a value of B. Then at a point (P, V) on the V vs P curve at B_*, where the drop shape is $Z(x)$, they satisfy

$$-B\psi = \frac{d}{dx} \frac{\psi_x}{(1 + Z_x^2)^{\frac{3}{2}}} + \lambda^2 \psi \tag{2.71}$$

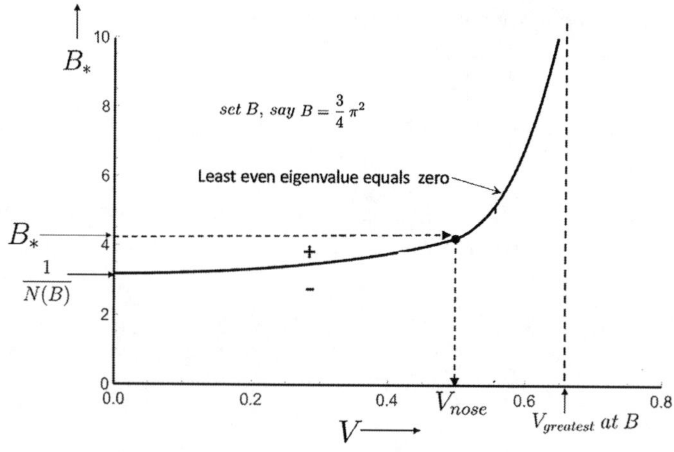

Fig. 2.17 Given $B \in \left(\frac{1}{4}\pi^2, \pi^2\right)$, and $B_* > \frac{1}{N(B)}$, the turning point curve, B_* vs V, gives us the value of V, viz., V_{nose}, where the least even eigenvalue first reaches zero upon advancing P along the V vs P curve

where $\psi = 0$ at $x = \pm 1$. This is the equation for the odd eigenvalues at the same drop shape, at the same volume, along the V vs P curve at $B_* = 0$. Thus if, upon advancing V upward from zero to $V_{greatest}(B)$ along the V vs P curve at $B_* = 0$, we find a volume, say $V(B)$, where the least odd eigenvalue has just reached zero, then so too at the volume $V(B)$ on the V vs P curve at any B_*, i.e., the odd eigenvalues are independent of B_*, they depend only on the drop shape.

Now, setting B_* to zero, we can find a value of B such that $V(B) = V_{greatest}(B)$. It divides the cases where $V(B)$ exists and where it does not. For all B's above about $\frac{1}{2}\pi^2$, we have a volume $V(B)$ where the least odd eigenvalue vanishes before V reaches its greatest value.

2.9 Locating Branch Points

To find where branch points lie, we integrate our drop shape equation to obtain

$$2(P - B_*V) + BV = \left.\frac{Z_x}{\sqrt{1 + Z_x^2}}\right|_{-1}^{1} \tag{2.72}$$

where due to $Z_x(x = 1) = -Z_x(x = -1)$ the right hand side is ordinarily not zero. Then we derive an equation for Z_x by differentiating our drop shape equation. It is

$$-BZ_x = \frac{d}{dx}\frac{(Z_x)_x}{(1 + Z_x^2)^{\frac{3}{2}}} \tag{2.73}$$

where at a branch point we notice that the odd eigenfunction corresponding to $\lambda^2 = 0$ satisfies the same equation, viz.,

$$-B\psi = \frac{d}{dx}\frac{\psi_x}{(1+Z_x^2)^{\frac{3}{2}}} \tag{2.74}$$

where $\psi = 0$ at $x = \pm 1$. Multiplying the first of these two equations by ψ, the second by Z_x, etc., we obtain

$$\int_{-1}^{1}\left\{Z_x\frac{d}{dx}\frac{\psi_x}{(1+Z_x^2)^{\frac{3}{2}}} - \psi\frac{d}{dx}\frac{(Z_x)_x}{(1+Z_x^2)^{\frac{3}{2}}}\right\}dx = 0 \tag{2.75}$$

whereupon we have

$$Z_x\frac{\psi_x}{(1+Z_x^2)^{\frac{3}{2}}}\Bigg|_{-1}^{1} = 0 \tag{2.76}$$

where Z is even, ψ is odd and $\psi_x(x = \pm 1)$ is not ordinarily zero. Thus we conclude that: at a branch point Z_x is zero at $x = \pm 1$, i.e., at a branch point the light fluid is just about to invade the heavy fluid.

The result then is that the straight line

$$2(P - B_* V) + BV = 0 \tag{2.77}$$

crosses the V vs P curve at a branch point. Having the V vs P curve at input values of B and B_* we can locate a branch point by drawing a straight line, cf., Fig. 2.18.

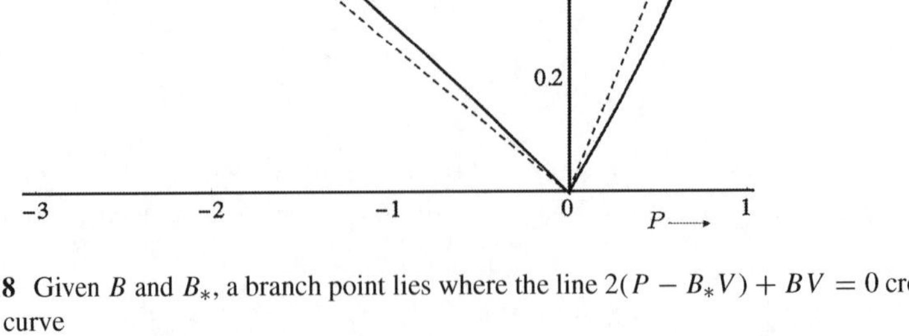

Fig. 2.18 Given B and B_*, a branch point lies where the line $2(P - B_* V) + BV = 0$ crosses the V vs P curve

2.10 Do Branch Points Come Before Turning Points?

Given the value of B, the points on a horizontal line at B_* in Fig. 2.19 are the points on the V vs P curve at B_* where $P = P_* + B_* V$ and where P_* depends on V at $B_* = 0$ and it is known.

Assume, now, that we are planning to run an experiment where we set the value of B, say, $B \in \left(\frac{1}{4}\pi^2, \pi^2\right)$ high enough that upon increasing V along the V vs P curve at $B_* = 0$ we reach the volume $V(B)$ before reaching the volume $V_{greatest}(B)$. Then we know the signs of the least odd and the least even eigenvalues along the V vs P curve, given B_*, cf., Fig. 2.20. This figure is drawn at B greater than $B_{\blacksquare}$, the Bond number at $B_* = 0$ at which the lowest even and odd eigenvalues reach zero together. At B less than $B_{\blacksquare}$ but greater than $\frac{1}{4}\pi^2$, the figure would look more or less similar but the vertical line drawn at $V(B)$ must be removed. At $B < B_{\blacksquare}$ the least odd eigenvalue remains positive for all V less than the greatest V as V increases to $V_{greatest}$.

At each B_* above $\frac{1}{N(B)}$ the V vs P curve starts out in the first quadrant. At first, the drops are stable. For each B_* below $\frac{1}{N(B)}$ the curves start out in the second quadrant where they remain and where the drops are unstable from the outset.

We expect to have V vs P curves of three kinds, each kind corresponding to one of the horizontal dotted lines in Fig. 2.21. We have marked the signs of the least odd and the least even eigenvalues along each line as V advances upward, given B_*. At any V, to find the input pressure:

$$V \implies P_* \text{ at } B_* = 0 \tag{2.78}$$

$$V, P_* \implies P = P_* + B_* V \text{ at } B_* \tag{2.79}$$

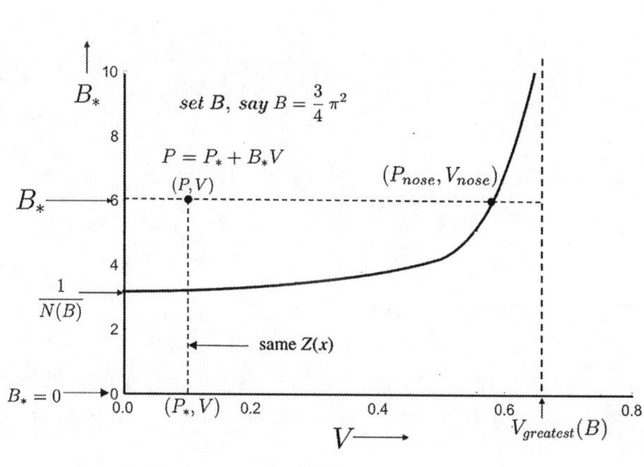

Fig. 2.19 The points on a horizontal line at B_* are the points on the V vs P curve

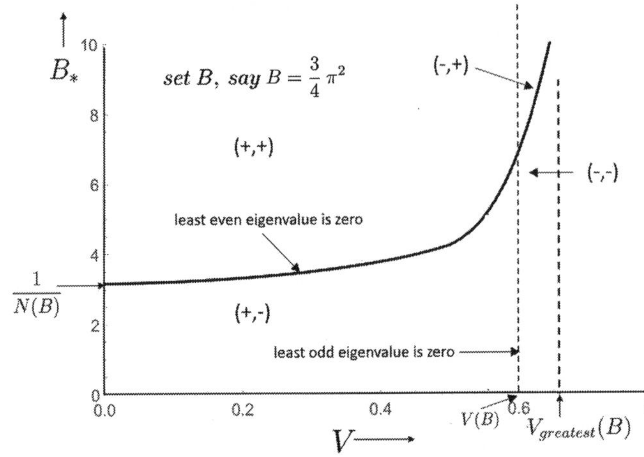

Fig. 2.20 Given B, $B > B_\blacksquare$ at $B_* = 0$, we have the signs of the least odd and the least even eigenvalues, (odd, even)

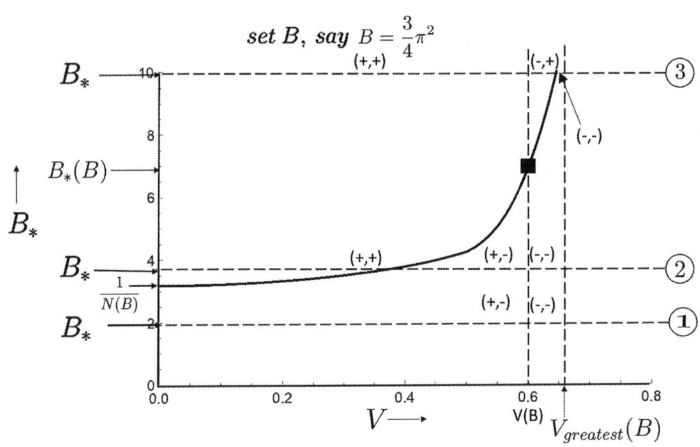

Fig. 2.21 The signs of the least odd and the least even eigenvalues along the V vs P curves at values of B_* marked ①, ② and ③, $B > B_\blacksquare$, at $B_* = 0$

Curve ① lies in the second quadrant. It is unstable and it cannot be seen in an experiment, cf., Fig. 2.22.

Curves ② and ③ start off stable in the first quadrant. Along curve ② a turning point appears before a branch point. Along curve ③ it is the reverse, cf., Figs. 2.23 and 2.24.

If a branch point comes before a turning point, eg., curve ③, drops under pressure control and under volume control break at the same $Z(x)$ and V and have the same pattern at breakup. The pressures differ by $B_* V$.

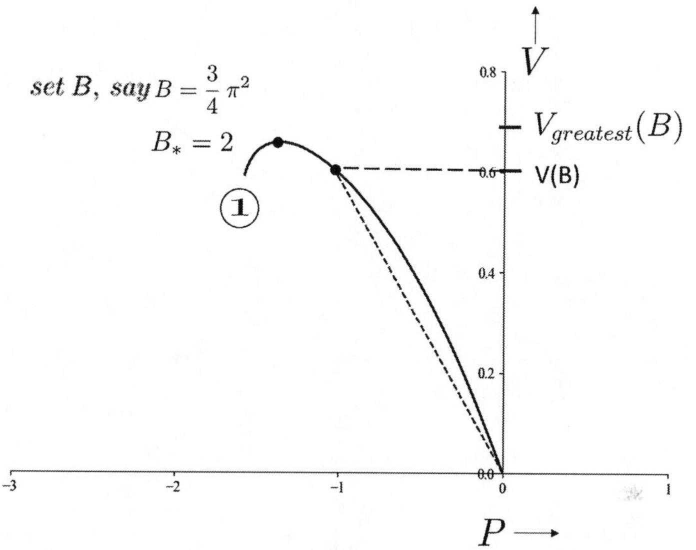

Fig. 2.22 Along curve ① the least even eigenvalue is negative

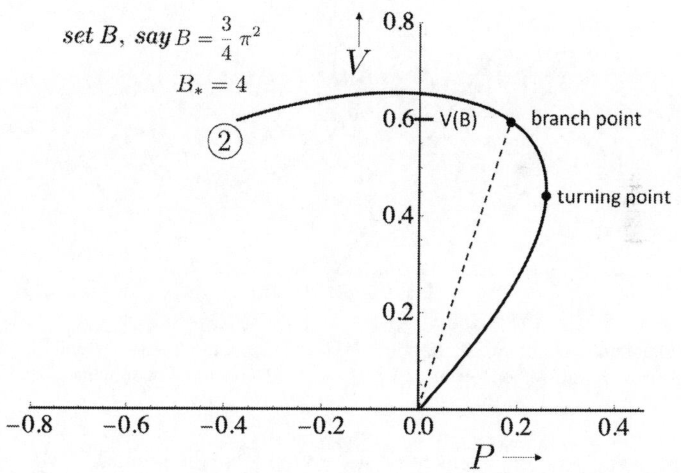

Fig. 2.23 Along curve ② a turning point comes before a branch point

The point marked ■ in Fig. 2.21 divides the values of B_* to which curves ② and ③ belong. Given B, it tells us $B_*(B)$ and $V(B)$ where the least odd and the least even eigenvalues reach zero together. It can be obtained by drawing two straight lines, given only the V vs P curve at $B_* = 0$, a volume control result.

The first line, $V = -\frac{2}{B}P$, locates $V(B)$. The second line, the tangent line at $V(B)$, tells us $B_*(B)$, viz., $B_*(B) = -\frac{1}{\textit{slope of the tangent}}$, cf., Fig. 2.25.

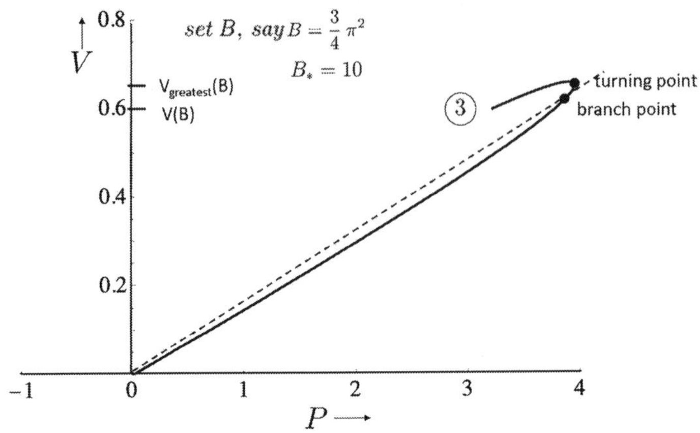

Fig. 2.24 Along curve ③ a branch point is reached before a turning point

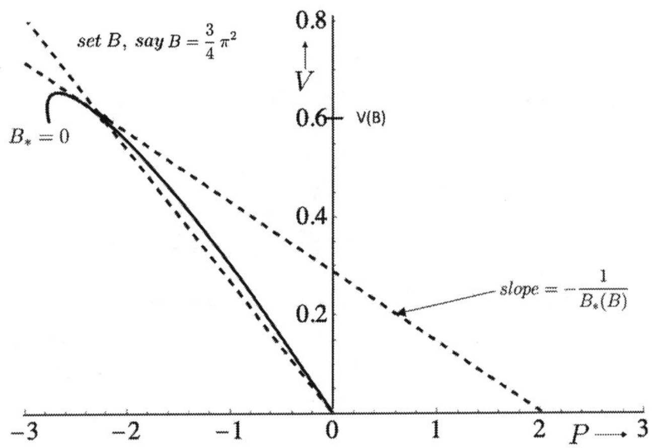

Fig. 2.25 Obtaining the point (V, B_*) in Fig. 2.21 where the least odd and the least even eigenvalues reach zero together. Given B, $V(B)$ and $B_*(B)$ result upon drawing two straight lines

No eigenvalue problem needs to be solved to locate the point ■ : $(V(B), B_*(B))$ where the least odd and the least even eigenvalues reach zero together.

2.11 Conclusion

No calculations need to be made in order to solve drop shape problems in the case where the pressure is the input variable. This assumes the problem has already been solved in the case where the volume is the input variable and that we have the V vs. P curve and the drop shapes along it for any value of B that we set. Drop shapes, critical drop shapes, turning points and branch points are all obtained simply by knowing what straight lines to draw. The only eigenvalues that arise are those equal to zero.

We obtain these results by assuming a simple one-dimensional model. But the connection of pressure-controlled experiments to volume-controlled experiments obtains for drops pinned to arbitrary closed curves where only drop shapes and turning points occupy our attention.

2.12 End Notes for Chap. 2

2.12.1 Solving the Eigenvalue Problem: Pressure Control

We set the value of B. Then we wish to solve the eigenvalue problem at a point (P, V) on the V vs. P curve at an input value of B_* where the drop shape is $Z(x)$. Our eigenvalue problem is

$$B_* \int_{-1}^{1} \psi dx - B\psi = \frac{d}{dx} \frac{\psi_x}{(1 + Z_x^2)^{\frac{3}{2}}} + \lambda^2 \psi \tag{2.80}$$

where $\psi = 0$ at $x = \pm 1$.

As B_* grows large, the integral of ψ must go to zero. Then, denoting the limiting product, $B_* \int_{-1}^{1} \psi dx$, by C, we have the eigenvalue problem for drops under volume control. Thus, as B_* grows large, the eigenvalues where the pressure is the control variable converge to the eigenvalues where the volume is the control variable.

At the point (P_*, V) on the V vs P curve at $B_* = 0$, where we have $P_* = P - B_* V$, we have the same $Z(x)$ and V, and we write our unconstrained eigenvalue problem, viz.,

$$- B\phi = \frac{d}{dx} \frac{\phi_x}{(1 + Z_x^2)^{\frac{3}{2}}} + \mu^2 \phi \tag{2.81}$$

where $\phi = 0$ at $x = \pm 1$. Its solutions are the solutions to the eigenvalue problem, Eq. (2.80), at $B_* = 0$. This eigenvalue problem came up in an end note to the drop shape problem under volume control. We assume we have its solutions at each point of the V vs P curve at $B_* = 0$ and we denote these solutions

$$\mu_1^2 < \mu_2^2 < \ldots$$

$$\phi_1, \phi_2, \ldots$$

$$even, \ odd, \ \ldots$$

The μ^2's decrease as V increases along the V vs P curve at $B_* = 0$. At $V = 0$, and $Z(x) = 0$, we have

$$\mu_1^2 = -B + \frac{1}{4}\pi^2 \tag{2.82}$$

and

$$\mu_2^2 = -B + \pi^2 \tag{2.83}$$

Our eigenvalue problem, viz., Eq. (2.80), has odd and even solutions. The odd solutions are $\mu_2^2, \phi_2(x); \mu_4^2, \phi_4(x); \ldots$ i.e., no matter the value of B_*, the odd solutions depend only on the drop shape $Z(x)$ and the drop shape is the same at B_*, P, V as it is at $B_* = 0, P_*, V$ so long as $P = P_* + B_* V$ where P_* is set by V at $B_* = 0$. Thus all (B_*, P)'s have the same odd eigenvalues as does $(B_* = 0, P_*)$, the values of $Z(x)$ and V remaining fixed.

To obtain the even solutions we expand the even ψ's in the even ϕ's. Thus having

$$\int_{-1}^{1} \phi_i^2 dx = 1 \tag{2.84}$$

we write

$$\psi = \sum_{i=1,3,\ldots} c_i \phi_i \tag{2.85}$$

where $c_i = \int_{-1}^{1} \psi\, \phi_i\, dx$. Then multiplying the ψ equation by ϕ_i, the ϕ_i equation by ψ, subtracting and integrating over $(-1, 1)$ we have, denoting $\int_{-1}^{1} \phi_i\, dx$ by I_i,

$$\frac{B_* I_i \int_{-1}^{1} \psi dx}{\lambda^2 - \mu_i^2} = c_i \tag{2.86}$$

Then multiplying by ϕ_i, summing and integrating we have

$$LHS \equiv \sum_{i=1,3,\ldots} \frac{I_i^2}{\lambda^2 - \mu_i^2} = \frac{1}{B_*} \tag{2.87}$$

This is our equation for the even eigenvalues. The even eigenfunctions obtain via Eq. (2.85), where the c_i's come from Eq. (2.86) at each even eigenvalue.

At the point (P_*, V) of the V vs P curve at $B_* = 0$, where the drop shape is $Z(x)$, we have the even eigenvalues. They are $\mu_1^2, \mu_3^2, \ldots$ and we can draw the curve LHS vs λ^2, cf., Fig. 2.26.

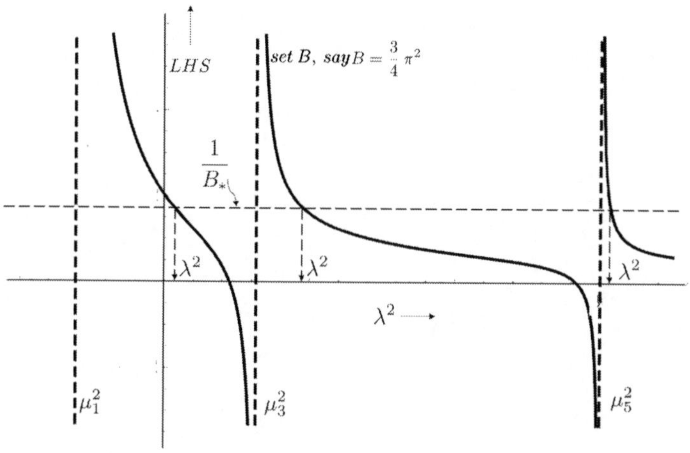

Fig. 2.26 Given B, at $Z(x)$ we have $\mu_1^2, \phi_1; \mu_3^2, \phi_3; \ldots$ and therefore LHS vs λ^2. At B_* we have the even eigenvalues marked by the down pointing arrows

Then upon setting the value of B_* we can obtain the even eigenvalues at the point (P, V) on the V vs P curve at B_* where the drop shape remains $Z(x)$ due to $P = P_* + B_* V$, cf., the down pointing arrows in Fig. 2.26. They lie on the intervals $(\mu_1^2, \mu_3^2), (\mu_3^2, \mu_5^2), \ldots$

Now we turn our attention to the least even eigenvalue. Our aim is to draw the curve along which it is zero. First, we derive its dependence on B_*, cf., Fig. 2.27. We plot the left most branch of the LHS vs λ^2 curve, having set B to a value on $\left(\frac{1}{4}\pi^2, \pi^2\right)$ and having obtained a drop shape $Z(x)$ by setting the value of V along the V vs P curve at $B_* = 0$. At $B_* = 0$ the least even eigenvalue is ${\mu_1}^2 < 0$. At large values of B_* it approaches the least even eigenvalues for a drop of shape $Z(x)$ under volume control. This is positive, cf., the up pointing arrow in Fig. 2.27.

Directing our attention to the least even eigenvalue, here is where we stand: at any $B \epsilon \left(\frac{1}{4}\pi^2, \pi^2\right)$ and any V along the V vs P curve at $B_* = 0$ we have a drop shape that is independent of B_*. This drop, unstable at $B_* = 0$, can be stabilized by increasing B_*. Thus under pressure control, a drop of a given shape, unstable at small B_*, becomes stable at large B_*. Increasing B_* is stabilizing. This drop would be stable under volume control.

On the other hand the least odd eigenvalue is independent of B_*, taking its $B_* = 0$ value for all B_*.

Now for each of the drop volumes marked in Fig. 2.28 we draw the leftmost branch of the LHS vs λ^2 curve in Fig. 2.29. Then at each V we obtain the value of B_* at which the least even eigenvalue reaches zero. In the figure the up pointing arrows mark the least even eigenvalue under volume control.

We can now plot the curve along which the least even eigenvalue is zero, cf., Fig. 2.30. This curve is the turning point curve we derived earlier before finding

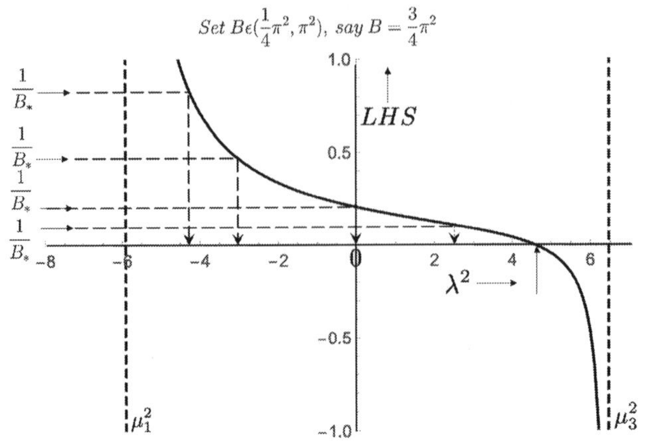

Fig. 2.27 Dependence of the even λ^2 with B_*

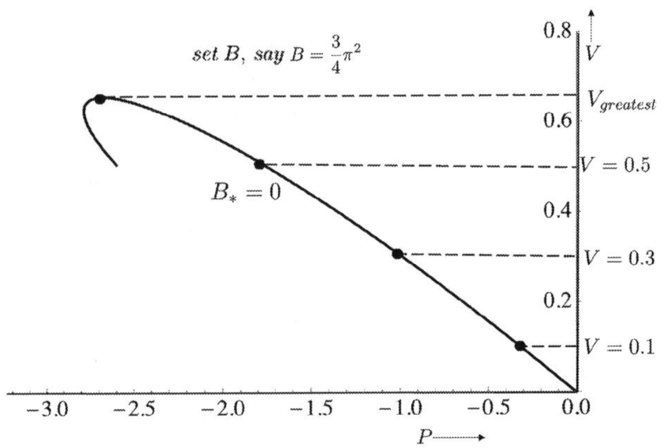

Fig. 2.28 The volumes, and therefore the drop shapes, at which we plot LHS vs λ^2 in Fig. 2.29

that it is also the least-even-eigenvalue-equal-to-zero curve. It is drawn at a value of $B \epsilon \left(\frac{1}{4}\pi^2, \pi^2\right)$, viz., $B = \frac{3}{4}\pi^2$.

We conclude this end note on eigenvalues by marking the signs of the least even eigenvalues along a V vs P curve at $B \epsilon \left(\frac{1}{4}\pi^2, \pi^2\right)$, first under volume control in Fig. 2.31 and then under pressure control in Fig. 2.32.

Thus, given $B \epsilon \left(\frac{1}{4}\pi^2, \pi^2\right)$, for $B_* > \frac{1}{N(B)}$ the V vs P curve starts out in the first quadrant and rises to a turning point. The least even eigenvalue starts out positive and falls to zero at the turning point, cf., the right hand curve in Fig. 2.32. For $B_* < \frac{1}{N(B)}$ the V vs P curve starts out in the second quadrant where it remains. The least even eigenvalue starts out negative and remains negative. At $B_* = 0$ it is ${\mu_1}^2$.

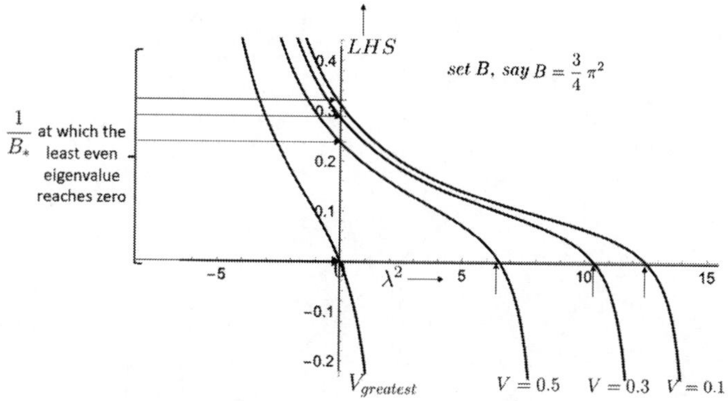

Fig. 2.29 Given $B \in \left(\frac{1}{4}\pi^2, \pi^2\right)$, we have the values of $\frac{1}{B_*}$ at which the least even eigenvalues reach zero at the volumes marked in Fig. 2.28. At $V_{greatest}$ we have $\frac{1}{B_*} = 0$

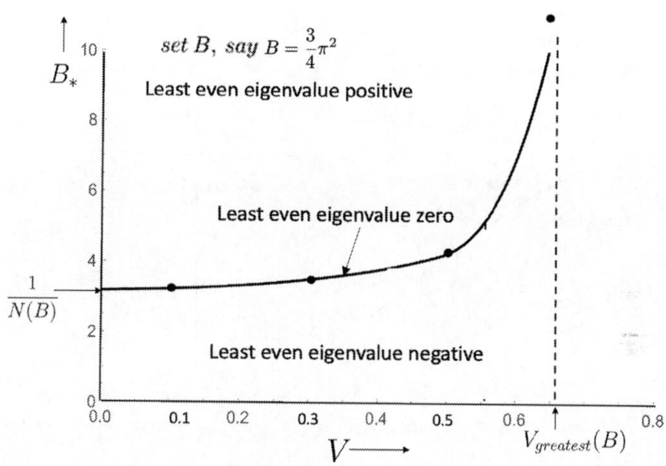

Fig. 2.30 Given $B \in \left(\frac{1}{4}\pi^2, \pi^2\right)$, then given $B_* > \frac{1}{N(B)}$ we have the volume at which the least even eigenvalue falls to zero along the V vs P curve, cf., Fig. 2.29

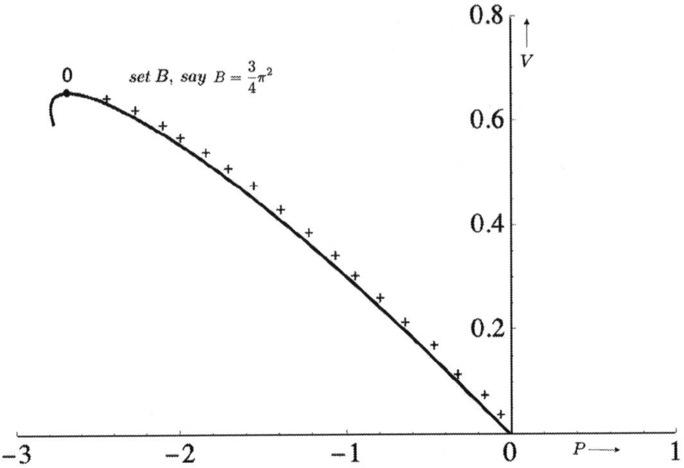

Fig. 2.31 Signs of the least even eigenvalues. Volume control. $B \in \left(\frac{1}{4}\pi^2, \pi^2\right)$

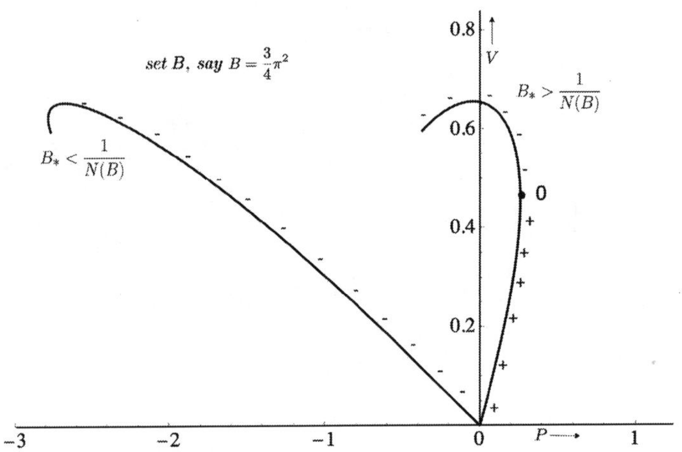

Fig. 2.32 Signs of the least even eigenvalues. Pressure control $B \in \left(\frac{1}{4}\pi^2, \pi^2\right)$

Chapter 3
The Stability of a Drop Pinned to an Arbitrary Plane Curve: Volume Control

Abstract This chapter generalizes the stability analysis to pendent drops pinned to arbitrary planar curves, moving beyond the circular geometry. While retaining volume control, the problem becomes inherently two-dimensional, with the geometry of the pinning boundary dictating the symmetry and structure of admissible drop shapes.

The governing equations for the drop shape are formulated, and the stability problem is again addressed through eigenvalue analysis. A key focus is on understanding how the symmetry of the boundary curve influences the symmetry and stability of the drop, and how the one-dimensional model developed earlier captures essential features of the more complex two-dimensional problem.

Special cases, including drops pinned to closed circular curves and sharp-edged geometries (e.g., rectangles), are examined to illustrate the broader applicability of the theory. The chapter establishes connections between the simplified one-dimensional framework and the full two-dimensional problem, showing that many qualitative features—such as the emergence of instability and the nature of critical points—are preserved.

The main conclusion is that the geometry of the pinning curve plays a fundamental role in determining stability, and that the methods developed in earlier chapters extend naturally to more general configurations, providing a unified approach to the analysis of pinned fluid interfaces.

3.1 Drop Pinned to a Circle

3.1.1 Introduction

We suppose now that our drop is pinned to a sharp-edged circle of radius R. We do this because there is a small surprise: having solved the one-dimensional model, we already know what to expect, but how it comes about is the surprise. A drop pinned to a circle is not a model problem, it is a possible experiment.

L. Johns, R. Narayanan, *The Shape and the Stability of Pendent Drops*,
Fluid Mechanics and Its Applications 144,
https://doi.org/10.1007/978-3-032-28202-6_3

3.1.2 A Few Formulas

Denoting the drop shape by $z = Z(r, \theta)$, we obtain its normal and its mean curvature via

$$\vec{n} = \frac{\vec{k} - \frac{Z_\theta}{r}\vec{i_\theta} - Z_r\vec{i_r}}{\left(1 + \frac{z_\theta^2}{r^2} + Z_r^2\right)^{\frac{1}{2}}} \tag{3.1}$$

and

$$2H = \frac{1}{\left(1 + \frac{z_\theta^2}{r^2} + Z_r^2\right)^{\frac{3}{2}}}\left\{\left(1 + \frac{Z_\theta^2}{r^2}\right)Z_{rr} - 2\frac{Z_r Z_\theta Z_{r\theta}}{r^2} + 2\frac{Z_r Z_\theta^2}{r^3} + \left(1 + Z_r^2\right)\left(\frac{Z_r}{r} + \frac{Z_{\theta\theta}}{r^2}\right)\right\} \tag{3.2}$$

Then, for the element of surface area and the drop volume we have

$$dA = \left(1 + \frac{Z_\theta^2}{r^2} + Z_r^2\right)^{\frac{1}{2}} r dr d\theta \tag{3.3}$$

and

$$V = -\int_0^{2\pi} d\theta \int_0^R Z(r, \theta)\, r dr \tag{3.4}$$

Assuming we have a symmetric drop shape, say, $Z_0(r)$, we introduce a small displacement, $\epsilon Z_1(r, \theta)$, and obtain formulas for $2H_0$ and $2H_1$. Upon writing

$$Z(r, \theta, \epsilon) = Z_0(r) + \epsilon Z_1(r, \theta) + \cdots \tag{3.5}$$

and

$$2H = 2H_0 + \epsilon 2H_1 + \cdots \tag{3.6}$$

where

$$2H_0 = 2H \mid_{\epsilon=0} \tag{3.7}$$

and

$$2H_1 = \frac{d}{d\epsilon} 2H \Big|_{\epsilon=0} \tag{3.8}$$

we have

$$2H_0 = \frac{1}{r}\frac{d}{dr}\frac{rZ_{0r}}{\left(1+Z_{0r}^2\right)^{\frac{1}{2}}} \tag{3.9}$$

and

$$2H_1 = \frac{1}{r}\frac{\partial}{\partial r}\frac{rZ_{1r}}{\left(1+Z_{0r}^2\right)^{\frac{3}{2}}} + \frac{1}{\left(1+Z_{0r}^2\right)^{\frac{1}{2}}}\frac{Z_{1\theta\theta}}{r^2} \tag{3.10}$$

3.1.3 *The Equation for the Drop Shape*

We direct our attention to the sketch in Fig. 1.1 where a heavy fluid lies above a light fluid and where the shape of the drop is the output, its volume the input, heavy fluid being added, and light fluid being removed. The drop shape is assumed to be symmetric. This corresponds to assuming the shape is even in the one-dimensional model.

Hydrostatic conditions prevail and the pressures in the heavy and light phases are

$$p = -\rho g z + \mathscr{C} \tag{3.11}$$

and

$$p^{\star} = -\rho^{\star} g z + \mathscr{C}^{\star} \tag{3.12}$$

The drop shape is then the solution to the Young-Laplace equation, viz.,

$$p^{\star} + \gamma 2H = p \tag{3.13}$$

whereupon to obtain $Z(r)$ we must solve

$$\mathscr{C} - \Delta\rho g\, Z = \gamma\, \frac{1}{r}\frac{d}{dr}\frac{rZ_r}{\left(1+Z_r^2\right)^{\frac{1}{2}}} \tag{3.14}$$

$$Z = 0 \;\; at \;\; r = R \tag{3.15}$$

and

$$2\pi \int_0^R Z(r) r dr = -V \tag{3.16}$$

where V is the input, $Z(r)$ and $\mathscr{C}$ are the outputs and where $\mathscr{C} = \mathscr{C} - \mathscr{C}^*$.

Now scaling r and $Z(r)$ by R, denoting $\frac{\Delta\rho g R^2}{\gamma}$ by B and writing P for $\frac{\mathscr{C} R}{\gamma}$, we have our scaled problem for the drop shape. It is

$$P - BZ = \frac{1}{r}\frac{d}{dr}\frac{rZ_r}{(1+Z_r^2)^{\frac{1}{2}}}, \quad Z = 0 \ at \ r = 1 \tag{3.17}$$

and

$$2\pi \int_0^1 rZdr = -V \tag{3.18}$$

At $V = 0$ we have $Z(r) = 0 = P$.

3.1.4 The Plan

Our plan is to set the value of B and then advance V upward from zero obtaining $Z(r)$ and P at each V along the way. Our aim is to obtain the critical value of V where the drop is just about to break.

To do this we solve an eigenvalue problem. At a volume V, where the drop shape is $Z(r)$, it is

$$C - B\psi = \frac{1}{r}\frac{\partial}{\partial r}\frac{r\psi_r}{(1+Z_r^2)^{\frac{3}{2}}} + \frac{1}{r^2}\frac{1}{(1+Z_r^2)^{\frac{1}{2}}}\psi_{\theta\theta} + \lambda^2\psi \tag{3.19}$$

$$\psi = 0 \ at \ r = 1 \tag{3.20}$$

and

$$\int_0^{2\pi} d\theta \int_0^1 r\psi dr = 0 \tag{3.21}$$

Then a drop of shape $Z(r)$ is stable if and only if we find that all the eigenvalues solving our eigenvalue problem are positive.

We set the value of B so that all the eigenvalues are positive at $V = 0$. The eigenvalues move to the left as V advances upward from $V = 0$. Thus upon advancing V our job is to find the value of V where the least eigenvalue first reaches zero. This value of V is the critical volume.

3.1.5 The Eigenvalue Problem

The eigenvalue problem has families of solutions denoted by $m = 0, m = 1, m = 2 \cdots$, where the solutions in each family are multiples of $\cos m\theta$. For $m = 1, 2, \cdots$ we have $\int_0^{2\pi} d\theta \int_0^1 \psi r dr = 0$ whereupon the constant C is zero. The least $m = 1$ eigenvalue is less than the least $m = 2, 3, 4 \cdots$ eigenvalues and we wish to know if the least eigenvalue first reaching zero is an $m = 0$ or $m = 1$ eigenvalue. The $m = 0$ and $m = 1$ eigenfunctions correspond to the even and odd eigenfunctions in our one-dimensional model problem.

At $m = 0$ our eigenvalue problem is

$$C - B\psi = \frac{1}{r}\frac{d}{dr}\frac{r\psi_r}{(1 + Z_r^2)^{\frac{3}{2}}} + \lambda^2\psi \tag{3.22}$$

$$\psi = 0 \ \ at \ \ r = 1 \tag{3.23}$$

and

$$\int_0^1 r\psi dr = 0 \tag{3.24}$$

At $m = 1$, upon setting $\psi = \hat{\psi}(r) \ cos \ \theta$, it is

$$-B\hat{\psi} = \frac{1}{r}\frac{d}{dr}\frac{r\hat{\psi}_r}{(1 + Z_r^2)^{\frac{3}{2}}} - \frac{1}{r^2}\frac{1}{(1 + Z_r^2)^{\frac{1}{2}}}\hat{\psi} + \lambda^2\hat{\psi} \tag{3.25}$$

and

$$\hat{\psi} = 0 \ \ at \ \ r = 1 \tag{3.26}$$

At $V = 0$, where $Z = 0$, we have : at $m = 0$: $\psi = A_0 J_0((\lambda^2 + B)^{\frac{1}{2}} \ r) + \frac{C}{\lambda^2 + B}$ and at $m = 1$: $\hat{\psi} = A_1 J_1((\lambda^2 + B)^{\frac{1}{2}} r)$. Thus the $m = 0$ eigenvalues satisfy

$$A_0 J_0((\lambda^2 + B)^{\frac{1}{2}}) + \frac{C}{\lambda^2 + B} = 0 \tag{3.27}$$

and

$$A_0 \int_0^1 r J_0((\lambda^2 + B)^{\frac{1}{2}} r) dr + \frac{1}{2}\frac{C}{\lambda^2 + B} = 0 \tag{3.28}$$

Then writing

$$\lambda^2 = -B + x^2 \tag{3.29}$$

we have

$$\frac{1}{x}J_1(x) - \frac{1}{2}J_0(x) = 0 \tag{3.30}$$

At $m = 1$ we have

$$J_1(x) = 0 \tag{3.31}$$

In both cases we have $\lambda^2 = 0$ where $B = x^2$. At $m = 0$, x is the least positive root of

$$\frac{1}{x}J_1(x) - \frac{1}{2}J_0(x) = 0 \tag{3.32}$$

whereas, at $m = 1$, x is the least positive root of $J_1(x) = 0$. In Fig. 3.1 we plot $J_1(x)$ and $J_1(x) - \frac{1}{2}x\ J_0(x)$ vs. x where we see that the least positive root of $J_1(x) = 0$ lies to the left of the least positive root of $J_1(x) - \frac{1}{2}x\ J_0(x) = 0$. Upon choosing B less than the square of the least positive root of $J_1(x) = 0$, i.e., less than about 15, we have all eigenvalues positive at $V = 0$.

Our plan then is to set B less than about 15, hold B fixed and increase V until we find the value of V where the least eigenvalue just reaches zero. This is the critical value of V at B, denoted V_{crit}, and we would like to know if the critical point is a branch point or a turning point, i.e., we would like to know if we can increase V and pass through V_{crit} or if have we reached the greatest possible value of V.

The critical point is a branch point if we can obtain $\overset{\bullet}{Z}$, where $\overset{\bullet}{Z} = \frac{dZ}{dV}$. The problem to be solved for $\overset{\bullet}{Z}$ obtains upon differentiating the drop shape problem with respect to V. It is

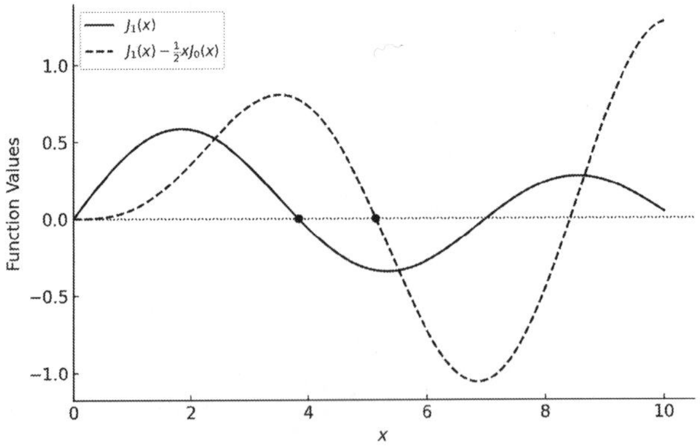

Fig. 3.1 Graphs of $J_1(x)$ and $J_1(x) - x\frac{1}{2}\ J_0(x)$

$$\dot{P} - B\dot{Z} = \frac{1}{r}\frac{d}{dr}\frac{r\dot{Z}_r}{(1+Z_r^2)^{\frac{3}{2}}} \tag{3.33}$$

$$\dot{Z} = 0 \ \ at \ \ r = 1 \tag{3.34}$$

and

$$\int_0^1 r\dot{Z}dr = -\frac{1}{2\pi} \tag{3.35}$$

This is a linear, non homogeneous equation for $\dot{Z}$. We wish to know if we can find a solution at $V = V_{crit}$.

Now if the first eigenvalue to reach zero occurs at $m = 0$ the eigenfunction satisfies

$$C - B\psi = \frac{1}{r}\frac{d}{dr}\frac{r\psi_r}{(1+Z_r^2)^{\frac{3}{2}}} \tag{3.36}$$

$$\psi = 0 \ \ at \ \ r = 1 \tag{3.37}$$

and

$$\int_0^1 r\psi dr = 0 \tag{3.38}$$

Thus to obtain $\dot{Z}$ a solvability condition must be satisfied. Upon multiplying the $\dot{Z}$ problem by $r\psi$, the ψ problem by $r\dot{Z}$, subtracting and integrating over $0 < r < 1$, we have the solvability condition. It is

$$C\int_0^1 r\dot{Z}dr = 0 \tag{3.39}$$

where

$$C\int_0^1 r\dot{Z}dr = -\frac{1}{2\pi}C \tag{3.40}$$

Because C is not zero at $m = 0$, solvability is not satisfied and $\dot{Z}$ does not exist. Thus the critical point must be a turning point if the least $m = 0$ eigenvalue is the first to reach zero. In this case we have reached the greatest possible value of V.

On the other hand the critical point may correspond to the least $m = 1$ eigenvalue reaching zero first. This being the case, all the eigenvalues solving the $m = 0$

eigenvalue problem, viz.,

$$C - B\psi = \frac{1}{r}\frac{d}{dr}\frac{r\psi_r}{(1+Z_r^2)^{\frac{3}{2}}} + \lambda^2\psi \tag{3.41}$$

$$\psi = 0 \ at \ r = 1 \tag{3.42}$$

and

$$\int_0^1 r\psi dr = 0 \tag{3.43}$$

must be positive, i.e., none are zero. Thus the homogeneous problem

$$\overset{\bullet}{P} - B\overset{\bullet}{Z} = \frac{1}{r}\frac{d}{dr}\frac{r\overset{\bullet}{Z}_r}{(1+Z_r^2)^{\frac{3}{2}}} \tag{3.44}$$

$$\overset{\bullet}{Z} = 0 \ at \ r = 1 \tag{3.45}$$

and

$$\int_0^1 r\overset{\bullet}{Z} dr = 0 \tag{3.46}$$

has only the solution $\overset{\bullet}{Z} = 0$. Then assuming that the Fredholm alternative holds for constrained problems we can conclude that $\overset{\bullet}{Z}$ is not zero whereupon we can advance V through V_{crit} and our critical point is a branch point.

Thus, upon advancing V upward from zero, the first time we find λ^2 equal to zero, either we have a branch point appearing before a turning point and we can continue increasing V or we have a turning point appearing before a branch point and we have reached the greatest value of V.

3.1.6 Do We Need to Solve Eigenvalue Problems?

Having set the value of B, we set a value of V, obtain $Z(r)$ and P and plot our results as a V vs P curve where we have the drop shape at each point along the curve. To identify critical points we notice that we can read turning points directly. Thus no eigenvalue problem needs to be solved to identify turning points.

So too in the case of branch points. To see why this is so, we first multiply the drop shape equation by $2\pi r$ and integrate over $0 < r < 1$ to obtain

$$\pi P + BV = 2\pi \left.\frac{Z_r}{(1+Z_r^2)^{\frac{1}{2}}}\right|_{r=1} \tag{3.47}$$

whereupon our aim is to establish the fact that drops at branch points have horizontal tangents where they contact the sharp edge, viz., we wish to obtain $Z_r = 0$ at $r = 1$. We begin by deriving an equation for Z_r. To do this we differentiate the drop shape equation with respect to r. Doing this we have

$$-BZ_r = \frac{1}{r}\frac{d}{dr}\frac{r\frac{d}{dr}Z_r}{(1+Z_r^2)^{\frac{3}{2}}} - \frac{1}{r^2}\frac{Z_r}{(1+Z_r^2)^{\frac{1}{2}}} \tag{3.48}$$

and we observe that the second term on the right-hand side has no counterpart in the one-dimensional model problem. It might block our attempt to derive $Z_r = 0$ at $r = 1$.

At a branch point, where the least λ^2 is zero at $m = 1$, $\hat{\psi}$ satisfies

$$-B\hat{\psi} = \frac{1}{r}\frac{d}{dr}\frac{r\hat{\psi}_r}{(1+Z_r^2)^{\frac{3}{2}}} - \frac{1}{r^2}\frac{\hat{\psi}}{(1+Z_r^2)^{\frac{1}{2}}} \tag{3.49}$$

and

$$\hat{\psi} = 0 \ \ at \ \ r = 1 \tag{3.50}$$

The second term on the right-hand side, coming from $\psi_{\theta\theta}$ at $m = 1$, turns out to be just what we need to learn that $Z_r = 0$ at $r = 1$. This is good luck and thus a surprise. We would not have had horizontal tangents at branch points had the least eigenvalue corresponding to $m = 2$ or 3 or $\cdots$ reached zero before the least eigenvalue corresponding to $m = 1$.

To learn that $Z_r = 0$ at $r = 1$ at a branch point, i.e., at an $m = 1$ critical point, we multiply the Z_r equation by $\hat{\psi}$, the $\hat{\psi}$ equation by Z_r, subtract and integrate over $0 < r < 1$ to obtain

$$-\left.\frac{Z_r r\hat{\psi}_r}{(1+Z_r^2)^{\frac{1}{2}}}\right|_{r=1} = 0 \tag{3.51}$$

and thus we have $Z_r|_{r=1} = 0$ assuming $\hat{\psi}_r$ is not zero at $r = 1$.

Having obtained $Z_r = 0$ at $r = 1$ at a branch point we turn to the integral of the drop shape equation. At a branch point it is

$$\pi P + BV = 0 \tag{3.52}$$

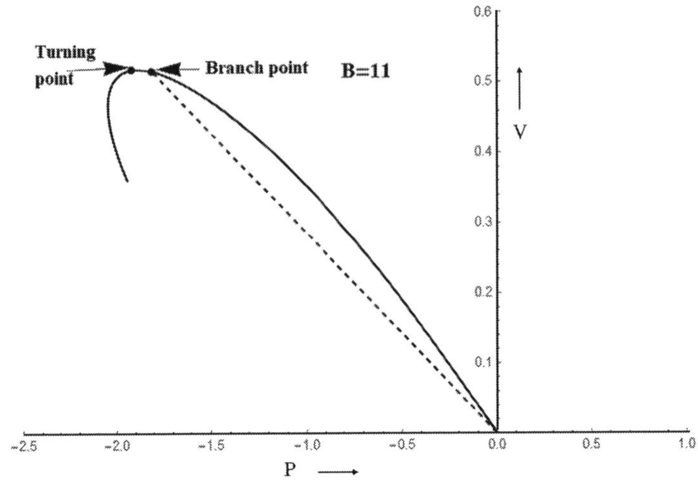

Fig. 3.2 Locating a branch point along the V vs P curve at $B = 11$ by drawing the straight line $\pi P + BV = 0$

Thus to identify a branch point along the V vs P curve, all we need to do is draw the straight line $\pi P + BV = 0$ and mark where it crosses the V vs. P curve. There we will find a branch point.

An example is sketched in Fig. 3.2 where B is large enough that a branch point appears before a turning point upon increasing V from zero.

3.1.7 *What Do the Two-dimensional Problem and the One-dimensional Model Have in Common?*

Here is what the drop pinned to a circle and the model problem have in common:

(1) The drop shape, $Z(r)$, being independent of θ, has the symmetry of the even solution $Z(x)$
(2) The $m = 0$ and $m = 1$ eigenfunctions have the symmetry of the even and odd eigenfunctions
(3) Branch points correspond to $m = 1$ or odd eigenvalues reaching zero first. Turning points correspond to $m = 0$ or even eigenvalues reaching zero first.
(4) Branch points correspond to horizontal tangents at the point of contact and can be found at the intersection of the straight line

$$2P + BV = 0 \tag{3.53}$$

or

$$\pi P + BV = 0 \tag{3.54}$$

and the curve V vs P.

Turning points can be found by inspecting the V vs. P curve.

(5) Branch points precede turning points at large values of B, turning points precede branch points at small values of B.

(6) The reader can derive the fact that V vs P curves start out in the first quadrant at small values of B and then work out the case where an outside pressure is the input variable.

We turn to a drop pinned to an arbitrary closed curve. No symmetry is assumed. Then we turn to a drop pinned to a rectangle restoring some of the symmetry we gave up in moving on from the circle.

3.2 Drops Pinned to a Closed Curve

We begin by assuming that our drop is pinned to a closed curve $\mathscr{C}$ lying in the x, y plane. It encloses a domain, $\mathcal{A}$, where $\mathbf{m}$ denotes the outward normal to $\mathscr{C}$, cf., Fig. 3.3.

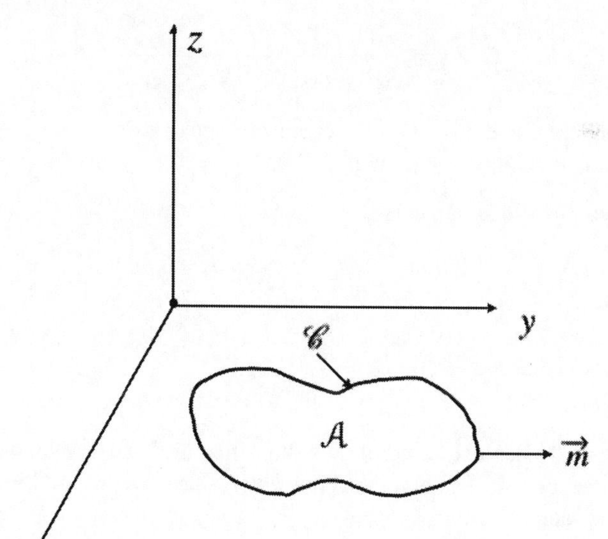

Fig. 3.3 An arbitrary closed curve lying in the x, y plane

The drop shape is denoted $z = Z(x, y)$, the input is the drop volume V where at each V we obtain $Z(x, y)$ and P by solving

$$P - BZ = \nabla \cdot \frac{\nabla Z}{(1 + \nabla Z \cdot \nabla Z)^{\frac{1}{2}}}, \quad Z = 0 \text{ on } \mathscr{C} \tag{3.55}$$

and

$$V = -\int_{\mathcal{A}} Z(x, y)\, dx\, dy \tag{3.56}$$

where P is set so that the volume constraint is satisfied. The variables Z, x, and y are scaled by a length L.

We set B and advance V upward from zero. At each V we have $Z(x, y)$ and P. To find out if a drop shape $Z(x, y)$ is stable to a displacement we solve an eigenvalue problem. It is

$$C - B\psi = \nabla \cdot \{\overset{\rightrightarrows}{\mathcal{L}}(Z) \cdot \nabla \psi\} + \lambda^2 \psi, \ \psi = 0 \ on \ \mathscr{C} \tag{3.57}$$

and

$$\int_{\mathcal{A}} \psi\, dx\, dy = 0 \tag{3.58}$$

where

$$\overset{\rightrightarrows}{\mathcal{L}}(Z) = \frac{\overset{\rightrightarrows}{I}(1 + \nabla Z \cdot \nabla Z) - \nabla Z \nabla Z}{(1 + \nabla Z \cdot \nabla Z)^{\frac{3}{2}}} \tag{3.59}$$

The drop is stable if and only if all the eigenvalues are positive.

To have stability at $V = 0$, where $Z(x, y) = 0$, we turn to our eigenvalue problem where $\overset{\rightrightarrows}{\mathcal{L}}(Z) = \overset{\rightrightarrows}{I}$. It is

$$C - B\psi = \nabla^2 \psi + \lambda^2 \psi, \ \psi = 0 \ on \ \mathscr{C} \tag{3.60}$$

and

$$\int_{\mathcal{A}} \psi\, dxdy = 0 \tag{3.61}$$

We aim to find the largest value of B so that for all smaller values of B all the eigenvalues are positive. To obtain the allowable values of B we multiply Eq. (3.60) by ψ and integrate over $\mathcal{A}$ to obtain

$$(B + \lambda^2) = \frac{\int_{\mathcal{A}} \nabla \psi \cdot \nabla \psi\, dxdy}{\int_{\mathcal{A}} \psi^2\, dxdy} \tag{3.62}$$

where the right hand side is greater than a least positive value, say, $B_{greatest}$. Thus for all B less than $B_{greatest}$ we have $\lambda^2 > 0$.

Thus we set B small enough that all eigenvalues are positive at $V = 0$. Then we advance V and at each V we have $Z(x, y)$ and P and we solve the eigenvalue problem directing our attention to the least eigenvalue. We advance V until the least eigenvalue reaches zero. This value of V is the critical drop volume.

Having obtained the critical drop volume we wish to know if our critical point is or is not a branch point. It is a branch point if we can obtain $\overset{\bullet}{Z}(x, y) = \frac{d\, Z(x,y)}{dV}$, where $\overset{\bullet}{Z}(x, y)$ solves the problem

$$\overset{\bullet}{P} - B\overset{\bullet}{Z} = \nabla \cdot \{\overset{\rightrightarrows}{\mathcal{L}} \cdot \nabla\overset{\bullet}{Z}\},\ \overset{\bullet}{Z} = 0 \text{ on } \mathscr{C} \tag{3.63}$$

and

$$1 = -\int_{\mathcal{A}} \overset{\bullet}{Z}(x, y)\, dx\, dy \tag{3.64}$$

At a critical point, where the least eigenvalue has just reached zero, we have

$$C - B\psi = \nabla \cdot \{\overset{\rightrightarrows}{\mathcal{L}} \cdot \nabla\psi\},\ \psi = 0 \text{ on } \mathscr{C} \tag{3.65}$$

and

$$\int_{\mathcal{A}} \psi\, dx\, dy = 0 \tag{3.66}$$

where ψ is not zero. Thus the homogeneous $\overset{\bullet}{Z}$ problem has a non zero solution,viz.,ψ, and therefore in order to obtain $\overset{\bullet}{Z}$ a solvability condition must be satisfied. The solvability condition is

$$C\int_{\mathcal{A}} \overset{\bullet}{Z}(x, y)\, dxdy = 0 \tag{3.67}$$

i.e, it is $C = 0$. Thus we have a branch point if and only if at our critical point we have C equal to zero.

To obtain C, in order to find out if it is or is not zero, we integrate the eigenvalue problem whereupon we have

$$CA = \int_{\mathcal{A}} dx\, dy\, \nabla \cdot \{\overset{\rightrightarrows}{\mathcal{L}} \cdot \nabla\psi\} \tag{3.68}$$

$$= \int_{\mathscr{C}} ds\, \vec{m} \cdot \{\overset{\rightrightarrows}{\mathcal{L}} \cdot \nabla\psi\} \tag{3.69}$$

then, setting

$$\vec{m} = \pm \frac{\nabla Z}{|\nabla Z|} \tag{3.70}$$

and using

$$\nabla Z \cdot \vec{\vec{\mathcal{L}}} = \frac{\nabla Z}{(1 + \nabla Z \cdot \nabla Z)^{\frac{3}{2}}} \tag{3.71}$$

we have

$$CA = \int_{\mathscr{C}} ds \frac{\pm \nabla Z}{|\nabla Z|} \frac{1}{(1 + \nabla Z \cdot \nabla Z)^{\frac{3}{2}}} \cdot \nabla \psi \tag{3.72}$$

$$= \int_{\mathscr{C}} ds \frac{\vec{m} \cdot \nabla \psi}{(1 + \nabla Z \cdot \nabla Z)^{\frac{3}{2}}} \tag{3.73}$$

Thus our critical point is a branch point if and only if

$$\int_{\mathscr{C}} ds \frac{\vec{m} \cdot \nabla \psi}{(1 + \nabla Z \cdot \nabla Z)^{\frac{3}{2}}} \tag{3.74}$$

is zero where $Z(x, y)$ denotes the drop shape at the critical point and ψ denotes the eigenfunction corresponding to the least λ^2 equal to zero.

Now, due to ψ=0 along $\mathscr{C}$, if $\nabla \psi$ is not zero at a point of $\mathscr{C}$, it will be a positive or negative multiple of $\vec{m}$. Thus obtaining a branch point will depend upon the symmetries of $\mathscr{C}$ and ψ. Ordinarily, unless $\mathscr{C}$ is special, we must assume that C is not zero and that our critical point is always a turning point, a point where V has reached its greatest value.

Here is where we stand: if our drop is pinned to a circle we have both branch points and turning points. We can identify where, along a V vs P curve, branch points and turning points lie by solving only the drop shape problem, i.e., we do not need to solve an eigenvalue problem. If our drop is pinned to an arbitrary closed curve we do not have branch points. Our critical points are always turning points and we do not need to solve an eigenvalue problem to find where they lie.

Now we turn our attention to rectangles, a geometry having less symmetry than circles but more than arbitrary curves.

But first we remind the readers that in the case of circles each of our eigenfunctions belongs to one of the symmetry classes, denoted $m = 0, m = 1, m = 2, \cdots$, and to obtain the least eigenvalue we turn our attention to the least eigenvalue in each class. Our straight line cuts the V vs. P curve at the $m = 1$ branch point, not at the $m = 2, \cdots$ branch points.

3.3 A Drop Pinned to a Sharp-edged Rectangle

Our drop is now pinned to a sharp-edged rectangle. In scaled variables our drop, having shape $z = Z(x, y)$, is defined for $-a < x < a$ and $-b < y < b$.

Our aim is to find out if drops pinned to rectangles have enough symmetry so that branch points can be found along V vs. P curves. We already know that there are turning points and that we do not need to solve eigenvalue problems in order to find out where they lie. Now a rectangle is a special case of an arbitrary closed curve and thus we have all we wish to know about rectangles by replacing the curve $\mathscr{C}$ by the perimeter of the rectangle marked ①, ②, ③ and ④ in Figs. 3.4 and 3.5

Again we increase V from zero where at each V we have $Z(x, y)$ and P. Then we obtain the least eigenvalue at each V and where the least eigenvalue reaches zero we have arrived at the critical volume. This critical point is a branch point if and only if the constant C at $\lambda^2 = 0$ is zero. To obtain C we must evaluate the right-hand side of the formula

$$CA = \int_{①+②+③+④} ds \, \frac{\vec{m} \cdot \nabla\psi}{(1 + \nabla Z \cdot \nabla Z)^{\frac{3}{2}}} \tag{3.75}$$

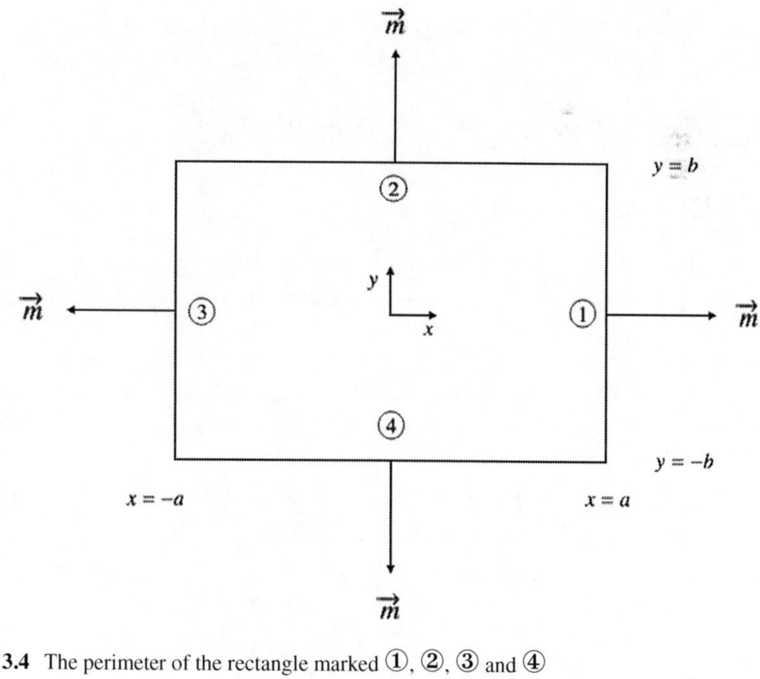

Fig. 3.4 The perimeter of the rectangle marked ①, ②, ③ and ④

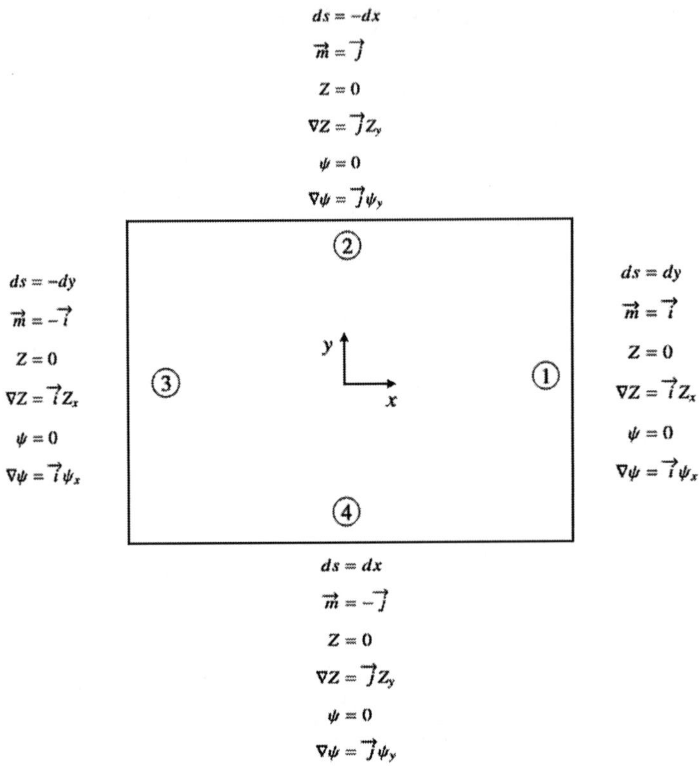

Fig. 3.5 Details needed to carry out the integration along the perimeter

Along ①, ②, ③ and ④ we have

$$① : \vec{i} \cdot \vec{\vec{\mathcal{L}}} = \frac{\vec{i}}{(1 + Z_x^2)^{\frac{3}{2}}} \quad \text{along} \quad x = a$$

$$② : \vec{j} \cdot \vec{\vec{\mathcal{L}}} = \frac{\vec{j}}{(1 + Z_y^2)^{\frac{3}{2}}} \quad \text{along} \quad y = b$$

$$③ : -\vec{i} \cdot \vec{\vec{\mathcal{L}}} = -\frac{\vec{i}}{(1 + Z_x^2)^{\frac{3}{2}}} \quad \text{along} \quad x = -a$$

$$\text{and } ④ : -\vec{j} \cdot \vec{\vec{\mathcal{L}}} = -\frac{\vec{j}}{(1 + Z_y^2)^{\frac{3}{2}}} \quad \text{along} \quad y = -b$$

and along ①, for example, our integral is

$$① = \int_{-b}^{b} dy \frac{\psi_x}{(1+Z_x^2)^{\frac{3}{2}}}\bigg|_{x=a} \tag{3.76}$$

Thus we have $\int_{perimeter} ds\ \vec{m} \cdot \frac{\nabla\psi}{(1+\nabla Z\cdot\nabla Z)^{\frac{3}{2}}}$ =①+③+②+④=

$$\int_{-b}^{b} dy \frac{\psi_x}{(1+Z_x^2)^{\frac{3}{2}}}\bigg|_{x=a} - \int_{-b}^{b} dy \frac{\psi_x}{(1+Z_x^2)^{\frac{3}{2}}}\bigg|_{x=-a}$$
$$+ \int_{-a}^{a} dx \frac{\psi_y}{(1+Z_y^2)^{\frac{3}{2}}}\bigg|_{y=b} - \int_{-a}^{a} dx \frac{\psi_y}{(1+Z_y^2)^{\frac{3}{2}}}\bigg|_{y=-b}$$

where along ③, i.e., along $x = -a$, s runs from 0 to $2b$, where $s = b - y$.

Now the drop shape has the symmetry:

$$\begin{aligned} Z = Z(x, y) &: \text{even in } x \text{ for all } y \\ &\quad \text{even in } y \text{ for all } x \\ Z_x = Z_x(x, y) &: \text{odd in } x \text{ for all } y \\ &\quad \text{even in } y \text{ for all } x \\ Z_y = Z_y(x, y) &: \text{even in } x \text{ for all } y \\ &\quad \text{odd in } y \text{ for all } x \end{aligned}$$

Therefore we have $Z_x^2(x = a, y) = Z_x^2(x = -a, y)$ and $Z_y^2(x, y = b) = Z_y^2(x, y = -b)$.

The eigenfunctions have one of four symmetries, denoted I, II, III or IV. In case I we have

$$\begin{aligned} \psi &: \text{odd in } x \text{ for all } y \\ &\quad \text{odd in } y \text{ for all } x \\ \psi_x &: \text{even in } x \text{ for all } y \\ &\quad \text{odd in } y \text{ for all } x \\ \psi_y &: \text{odd in } x \text{ for all } y \\ &\quad \text{even in } y \text{ for all } x \end{aligned}$$

then, due to $\psi_x(x = a, y) = \psi_x(x = -a, y)$ and $\psi_y(x, y = b) = \psi_y(x, y = -b)$, we have ①+③= 0, ②+④= 0 and therefore $C = 0$.

In case II we have

$$\begin{aligned}\psi &: \text{even in } x \text{ for all } y\\ &\quad \text{odd in } y \text{ for all } x\\ \psi_x &: \text{odd in } x \text{ for all y}\\ &\quad \text{odd in } y \text{ for all } x\\ \psi_y &: \text{even in } x \text{ for all } y\\ &\quad \text{even in } y \text{ for all } x\end{aligned}$$

then, due to $\psi_y(x, y = b) = \psi_y(x, y = -b)$, we have ②+④ = 0. And, due to $\psi_x(x = -a, y) = -\psi_x(x = a, y)$, we have

$$①+③ = 2\int_{-b}^{b} dy \left.\frac{\psi_x}{(1 + Z_x^2)^{\frac{3}{2}}}\right|_{x=a} \tag{3.77}$$

and because ψ_x is odd in y, ① + ③ = 0 and in case II we have $C = 0$.

Likewise in case III where

$$\begin{aligned}\psi &: \text{odd in } x \text{ for all } y\\ &\quad \text{even in } y \text{ for all } x\\ \psi_x &: \text{even in } x \text{ for all } y\\ &\quad \text{even in } y \text{ for all } x\\ \psi_y &: \text{odd in } x \text{ for all } y\\ &\quad \text{odd in } y \text{ for all } x\end{aligned}$$

we obtain $C = 0$. There remains case IV where,

$$\begin{aligned}\psi &: \text{even in } x \text{ for all } y\\ &\quad \text{even in } y \text{ for all } x\\ \psi_x &: \text{odd in } x \text{ for all } y\\ &\quad \text{even in } y \text{ for all } x\\ \psi_y &: \text{even in } x \text{ for all } y\\ &\quad \text{odd in } y \text{ for all } x\end{aligned}$$

whereupon we have

$$①+③=\int_{-b}^{b}dy\frac{\psi_x}{(1+Z_x^2)^{\frac{3}{2}}}\bigg|_{x=a} \tag{3.78}$$

where ψ_x is even in y and

$$②+④=\int_{-a}^{a}dx\frac{\psi_y}{(1+Z_y^2)^{\frac{3}{2}}}\bigg|_{y=b} \tag{3.79}$$

where ψ_y is even in x. Ordinarily neither ① + ③ nor ② + ④ is zero, nor is ① + ② + ③ + ④ zero. Thus in case IV we assume C is not zero.

We imagine, then, that upon increasing V and advancing along the V vs. P curve, we will find critical points where the least eigenvalue reaches zero for each of the four symmetry cases. Three of these critical points will be branch points, one will be a turning point. The turning point will be at the greatest drop volume. One or more or possibly none of the branch points may appear before the greatest volume is reached.

3.4 Conclusion

Here is what we have. By turning from circles to rectangles we do not lose branch points, though we do lose branch points by going all the way to arbitrary closed curves. In the case of circles, we have branch points at $m = 1, 2, 3, \cdots$ The branch point at $m = 1$ appears first on increasing V from zero and it can be obtained by drawing the straight line

$$PA + BV = 0 \tag{3.80}$$

Our question, then, is this: in the case of a rectangle upon drawing the straight line

$$PA + BV = 0 \tag{3.81}$$

do we locate a branch point along a V vs. P curve?

To obtain $PA + BV$ we integrate the drop shape equation. Doing this, assuming the light fluid has not invaded the heavy fluid, we have, at a point (P, V) on the V vs P curve where the drop shape is $Z(x, y)$,

$$PA + BV = \int_A dA\,\nabla\cdot\frac{\nabla Z}{(1+\nabla Z\cdot\nabla Z)^{\frac{1}{2}}} \tag{3.82}$$

$$= \int_{\mathscr{C}} ds\, \vec{m} \cdot \frac{\nabla Z}{(1 + \nabla Z \cdot \nabla Z)^{\frac{1}{2}}} \tag{3.83}$$

$$= \int_{①+②+③+④} ds\, \frac{|\nabla Z|}{(1 + \nabla Z \cdot \nabla Z)^{\frac{1}{2}}} \tag{3.84}$$

Thus in order to have $PA + BV = 0$ at a branch point we need to have a drop shape where $|\nabla Z|$ is zero all along ①+②+③+④. As unlikely as this might be, for it to be possible we need Z_x zero along ① and ③ and Z_y zero along ② and ④.

FSC
www.fsc.org
MIX
Papier | Fördert
gute Waldnutzung
FSC® C083411

Kolibri 360 GmbH
Ferdinand-Jühlke-Straße 7
99095 Erfurt, Deutschland
produktsicherheit@kolibri360.de